A Survey of Aircraft Structural-Life Management Programs in the U.S. Navy, the Canadian Forces, and the U.S. Air Force

Yool Kim, Stephen Sheehy, Darryl Lenhardt

Prepared for the United States Air Force

PROJECT AIR FORCE

The research described in this report was sponsored by the United States Air Force under Contract F49642-01-C-0003. Further information may be obtained from the Strategic Planning Division, Directorate of Plans, Hq USAF.

Library of Congress Cataloging-in-Publication Data

Kim, Yool.
 A survey of aircraft structural life management programs in the U.S. Navy, the Canadian forces, and the U.S. Air Force / Yool Kim, Stephen Sheehy, Darryl Lenhardt.
  p. cm.
 Includes bibliographical references.
 "MG-370."
 ISBN 0-8330-3862-1 (pbk. : alk. paper)
 1. Airplanes, Military—United States—Maintenance and repair—Management.
2. Airplanes, Military—Canada—Maintenance and repair—Management.
I. Sheehy, Stephen. II. Lenhardt, Darryl. III. Title.

UG1243.K493 2006
623.74'60288—dc22

                                              2005028956

The RAND Corporation is a nonprofit research organization providing objective analysis and effective solutions that address the challenges facing the public and private sectors around the world. RAND's publications do not necessarily reflect the opinions of its research clients and sponsors.

RAND® is a registered trademark.

Published 2006 by the RAND Corporation
1776 Main Street, P.O. Box 2138, Santa Monica, CA 90407-2138
1200 South Hayes Street, Arlington, VA 22202-5050
4570 Fifth Avenue, Suite 600, Pittsburgh, PA 15213
RAND URL: http://www.rand.org/
To order RAND documents or to obtain additional information, contact
Distribution Services: Telephone: (310) 451-7002;
Fax: (310) 451-6915; Email: order@rand.org

Preface

The Air Force's Aircraft Structural Integrity Program (ASIP) plays a critical role in fleet management maintaining airworthiness because it provides a framework for how the Air Force sustains its airframes. This report describes the findings from a survey of aircraft structural-life management programs in the U.S. Air Force, the U.S. Navy, and the Canadian Forces. The purpose of the study is to identify potential opportunities for the Air Force to improve its ASIP.

This work was performed under a project entitled "Understanding and Addressing the Effects of Aging Aircraft," which was sponsored by the Air Force Directorate of Maintenance (AF/ILM) and the Air Force Directorate of Strategic Planning (AF/XPX) and was conducted within the Resource Management Program of RAND Project AIR FORCE. This research is intended to be of interest to those involved in aircraft structural-life management and fleet management in the Air Force.

Previous RAND work on aging aircraft and fleet management issues has included research on aircraft maintenance workload forecasting, aircraft availability forecasting, and development of a fleet assessment methodology, as documented in

- Raymond A. Pyles, *Aging Aircraft: USAF Workload and Material Consumption Life Cycle Patterns*, Santa Monica, Calif.: RAND Corporation, MR-1641-AF, 2003.
- Edward G. Keating and Matthew Dixon, *Investigating Optimal Replacement of Aging Air Force Systems*, Santa Monica, Calif.: RAND Corporation, MR-1763-AF, 2003.

RAND Project AIR FORCE

RAND Project AIR FORCE (PAF), a division of the RAND Corporation, is the U.S. Air Force's federally funded research and development center for studies and analyses. PAF provides the Air Force with independent analyses of policy alternatives affecting the development, employment, combat readiness, and support of current and future aerospace forces. Research is conducted in four programs: Aerospace Force Development; Manpower, Personnel, and Training; Resource Management; and Strategy and Doctrine.

Additional information about PAF is available on our Web site at http://www.rand.org/paf.

Contents

Figures

Tables

Summary

The U.S. Air Force owns and operates approximately 6,000 aircraft to meet its force requirements. The average age of these aircraft is approximately 22 years and is expected to continue to rise. Many of the older aircraft are facing aging issues, such as structural deterioration due to fatigue, and many aircraft are expected to encounter such issues as the Air Force plans to keep aging aircraft in service for many years.

Fatigue is a process in which damage accumulates in material subjected to alternating or cyclic loading. This damage may culminate in cracks, which will eventually lead to complete fracture after a sufficient number of load cycles. Concern is growing in the Air Force that structural deterioration in aging aircraft is increasing the maintenance workload, reducing aircraft readiness, and potentially increasing safety risks (Pyles, 2003).

Since 1958, the Air Force has relied on its Aircraft Structural Integrity Program (ASIP) to achieve and maintain the structural safety of its aircraft. ASIP provides a framework for establishing and sustaining structural integrity throughout the aircraft's life.[1] The program's overarching objective is to prevent structural failures and to do so cost-effectively and without losing mission capability. ASIP is a key contributor to the Air Force's force management processes, and the

---

[1] Note that the term *ASIP* applies both to the overall program of the service (and of the Canadian Forces) and to individual programs tailored for particular aircraft types. Each individual program would include the aircraft designator in its name (e.g., the C-130 ASIP).

program's ongoing viability will be critical as the Air Force continues to operate an aging force to meet operational needs.

In recent years, some issues have been raised about inadequate implementation of ASIP. The concern is that an aging force, budget pressures, diminishing program regulation, and challenges in communicating structural condition and structural needs to decisionmakers may be leading to omission or incomplete performance of ASIP tasks. (See pp. 4–6.)

A further concern is that these factors may result in loss of control of ASIP, lack of visibility into the structural conditions of aircraft, and resource-allocation challenges for ASIP. The effectiveness of ASIP could be degraded, which would adversely affect operational effectiveness, flight safety, and fleet sustainment costs.

This report surveys aircraft structural-life management programs in the U.S. Navy, the Canadian Forces, and the U.S. Air Force to offer insights into how the Air Force could strengthen ASIP, particularly in enabling (1) independent and balanced regulation, (2) clear and timely communications, and (3) adequate and stable resources to achieve ASIP effectiveness. Table S.1 compares the technical and operational backgrounds for each service, and Table S.2 summarizes the key characteristics of each program.

The U.S. Navy's Aircraft Structural-Life Management

The Navy operates approximately 2,000 aircraft, based both on carriers and on land. In part because of the limited space and facilities on carriers for inspection and repairs, the Navy takes a "safe life" approach to structural-life management. Under this approach, airframes are assumed to be "flawless" at the time of manufacture, and aircraft are retired by the time fatigue cracks in the airframe initiate, which the Navy defines as reaching a length of 0.01 inch.

The Navy has an explicit policy on structural-life management. It establishes strict structural-life limits for each aircraft type and, to ensure structural safety, requires that aircraft not exceed these limits.

Table S.1
Comparison of Technical Basis and Operational Factors

	U.S. Navy	Canadian Forces	U.S. Air Force
Force structure	About 2,000 fixed-wing aircraft of 20 types	About 350 fixed-wing aircraft of 12 types	About 6,000 fixed-wing aircraft of 40 types
Operational environment	Carrier- and land-based	Land-based	Land-based
Technical basis	Safe life	Mix of safe life, damage tolerance	Damage tolerance

These limits are established based on the fatigue-life limit of the airframe and its critical components.

To ensure that the aircraft do not exceed their fatigue-life limits during service, the Navy tracks individual aircraft fatigue life in terms of a standard quantifiable metric, fatigue-life expended (FLE). A centralized program rigorously tracks the FLE for all aircraft and disseminates the information in a formal report to the organizations that support and operate the aircraft. Rigorous and accurate monitoring of fatigue life is critical to the Navy because, under the safe-life approach, there is no routine inspection for cracks to validate the structural condition. The centralized fatigue-life tracking program further has a dedicated funding line to provide independence to its assessments and to ensure that this critical task is carried out.

The Program Manager for Air (PMA) is responsible for the total life-cycle management of the designated fleet. PMA has the ownership and decision authority for structural-life management of the fleet (except for fatigue-life tracking). PMA uses the FLE information in making resource allocation decisions, such as conducting a service-life extension program, force structure planning, and scheduling modifications. The Naval Air System Command's Structures Division has regulatory responsibility for the technical aspects of structural-life management, providing an independent technical assessment of PMA's structural-life management decisions. The communications between the principal organizations involved in structural-life management are primarily informal, facilitated by their working relationship and colocation.

Table S.2
Summary of Key Characteristics in the U.S. Navy, the Canadian Forces, and the U.S. Air Force's Aircraft Structural-Life Management Programs

U.S. Navy	• Structural-life management policies are explicit.
	• Authority for the technical aspects of structural-life management is centralized.
	• A single, standard metric, FLE, conveys structural conditions.
	• Results of rigorous fatigue-life tracking are disseminated frequently, through a formal fatigue-life report.
	• Close working relationships and colocation promote and facilitate informal communication.
	• The structural-life monitoring program has dedicated funding.
Canadian Forces	• The policy is broad and is based on the concept of airworthiness.
	• The regulatory structure is independent but organizationally centralized.
	• Regulations exist to ensure communication and sharing of critical information.
	• Colocation and close working relationships facilitate informal communication.
	• A single authority (the weapon system manager [WSM]) controls funding for structural-life management for the designated fleet.
	• The ASIP master plan provides formal planning for resource management.
U.S. Air Force	• Policies are broad and flexible and are based on broad objectives.
	• The regulatory structure is flexible and decentralized, with minimal regulation and oversight.
	• Visibility of ASIPs and structural conditions across a command is limited.
	• Communications with the lead command on ASIP and structural issues are limited.
	• A single authority (the lead command) controls funding for structural-life management of the multiple fleets in the command.

The Canadian Forces' Aircraft Structural-Life Management

The Canadian Forces operate approximately 350 land-based aircraft of 12 different aircraft types. Because they are based on U.S. Navy designs, the Canadian Forces had originally implemented a safe-life approach to structural-life management. As the Canadian Forces have sought to extend the service lives of their aircraft, however, they have shifted to a "damage tolerance" approach. Damage tolerance assumes that the material of the airframe has flaws at the time of manufacture and that slowly growing cracks in the structure can be tolerated until they are detected and repaired. Structural inspection intervals are determined to ensure that a crack does not reach its critical size without being detected. Unlike the U.S. Navy, the Canadian Forces do not have carrier-based aircraft; thus, implementing a routine inspection for cracks to accommodate the damage-tolerance approach was not restricted by the space limitations on carriers.

The Canadian Forces take a regulatory approach to structural-life management. An independent regulatory authority, the Technical Airworthiness Authority (TAA), provides regulations and oversight for all weapon systems' ASIPs and assesses compliance. The governing policy regarding structural integrity is broad and is based on the concept of "airworthiness," in which the airworthiness requirements are defined in each aircraft type's basis of certification. An aircraft must remain in compliance with its basis of certification throughout its service life to be considered "airworthy."

The broad regulations allow each WSM to customize an ASIP for the specific weapon system. The TAA evaluates ASIP compliance on a case-by-case basis via formal airworthiness monitoring and approval processes. The approval processes focus on the tasks that are the linchpins of aircraft structural-life management, such as the airworthiness certification and design-change certification processes, to balance the level of regulation.

The regulatory approach requires considerable formal communication. The formal processes require documentation of critical information for traceability and planning purposes, as well as for assessing compliance. Additionally, colocation of key authorities in structural-life management facilitates informal communication among them.

The resource management plan is formally documented in the weapon system's ASIP master plan. As the funding and decision authority for the fleet's ASIP, the WSM must approve the plan. The master plan includes the short- and long-term tasks necessary for maintaining the structural integrity for the fleet. TAA's regulatory role provides independent assessments of WSM's resource allocation decisions.

The U.S. Air Force Aircraft Structural-Life Management

The U.S. Air Force operates a much-larger force with a wider range of aircraft types, about 40. The Air Force's ASIP is based on the damage-tolerance philosophy described earlier. The governing policy on ASIP is broad, focusing on the program's objectives, to allow tailoring of an ASIP for each aircraft type. Each system program director (SPD) is responsible for implementing an ASIP for its fleet. The lead command has the funding authority for the fleet management of the multiple fleets within the command, including ASIPs. As a result, the lead command has a significant influence on ASIP implementation.[2]

The Air Force has a flexible, decentralized regulatory structure with minimal ASIP regulation and oversight. There is no regulation to enforce certain ASIP tasks or to provide an independent technical assessment of the lead command's decisions about structural integrity. Because of this broad policy, the U.S. Air Force recommends multiple measures for assessing ASIP compliance, including mishap rates due to structural failure. This measure, however, is problematic because it is a lagging indicator.

The Air Force does not have a standard metric for communicating aircraft structural condition, partly because of the decentralized ASIP implementation. As a result, the lead command has a limited commandwide view of the structural condition of its fleets, and this makes understanding the relative states of the fleets for resource-allocation purposes challenging.

Communication between the SPD and the lead command about ASIP and structural condition is limited because of the limited involvement of the lead command in the ASIP process (other than

---

[2] When more than one major command possesses the same type of weapon system, one of them will be designated as the *lead command* for that system.

budget programming) and the geographic separation between them. Some lead commands use a technical liaison to facilitate the communication with the SPDs and to better understand the implications of the resource decisions (e.g., risk of structural failure, effects on operational effectiveness, preventing costly repairs).

Observations About Different Approaches

Explicit policy on ASIP provides clarity on ASIP compliance but limits flexibility in structural-life management. Broad policy on ASIP, on the other hand, enables flexibility in ASIP implementation for tailoring but risks lack of clarity about what constitutes acceptable ASIP compliance. The policy should be sufficiently explicit to provide general guidance on ASIP compliance but should rely on independent assessment of ASIP compliance on a case-by-case basis to enable tailoring. (See pp. 67–68.)

ASIP regulations can provide checks and balances for structural-life management, enable clear and timely communication, and promote stable and adequate resources for ASIP. Regulations could also lead to complex processes and management inefficiencies. The regulations should thus focus on elements of ASIP that are critical to the program's viability to ensure a balance between its control and flexibility. (See pp. 68–69, 72–75.)

Centralization enables standardization of program management and a forcewide view of ASIP compliance and aircraft fleet status, while decentralization enables tailoring to a specific weapon system to achieve a cost-effective ASIP. Centralization of a set of selective ASIP tasks, where standardization is useful, could still allow other aspects of ASIP to be tailored for cost-effectiveness. (See p. 71.)

Regulations, communications, and resource-management approaches are highly interdependent and need to complement each other and the context of the program (e.g., safe-life versus damage tolerance) to achieve ASIP effectiveness. Operational factors, such as the force size, may present certain scalability challenges. The U.S. Air Force's large-scale force with its wide range of aircraft types may pose

Abbreviations

ACC	Air Combat Command
AETC	Air Education and Training Command
AFB	Air Force base
AFFVB	Air Force Fleet Viability Board
AFI	Air Force Instruction
AFMC	Air Force Materiel Command
AFPD	Air Force policy directive
AFR	Air Force regulation
AFSOC	Air Force Special Operations Command
AIA	Airworthiness Investigative Authority
ALC	Air Logistics Center
ALC/EN	Air Logistics Center, Engineering Directorate
ALEX	Aircraft Life Extension
AMC	Air Mobility Command
ARB	Airworthiness Review Board
ASC	Aeronautical Systems Center
ASC/EN	Aeronautical Systems Center, Engineering Directorate
ASIP	Aircraft Structural Integrity Program
ASLS	Aircraft Structural Life Surveillance
CNO	Chief of Naval Operations

DoD	Department of Defense
FAA	Federal Aviation Administration
FLE	fatigue life expended
FLEI	fatigue life expended index
FLI	fatigue life index
FSMP	Force Structural Maintenance Plan
FST	fleet support team
IAT	Individual Aircraft Tracking
L/ESS	Loads and Environmental Spectra Survey
MAJCOM	major command
MDS	mission, design, series
MIL-HDBK	military handbook
MIL-STD	military standard
NAVAIR	Naval Air Systems Command
NAVAIRINST	NAVAIR instruction
NRC	National Research Council
OAA	Operational Airworthiness Authority
OPNAV	Office of the Chief of Naval Operations
OPNAVINST	OPNAV instruction
OPR	Office of Primary Responsibility
PEO	program executive officer
PMA	Program Manager for Air
POM	Program Objective Memorandum
SAF/AQ	Office of the Assistant Secretary of the Air Force for Acquisition
SAFE	Structural Assessment of Fatigue Effects
SLAP	Structural Life Assessment Program
SLEP	Service-Life Extension Program
SM	single manager

SPD	system program director
SPO	system program office
TAA	Technical Airworthiness Authority
T/M/S	type/model/series
TYCOM	type command
WSM	weapon system manager

CHAPTER ONE

Introduction

The Air Force's Aging Fleets

The U.S. Air Force currently owns and operates approximately 6,000 aircraft to meet its force requirements. Because of budget pressures and high replacement costs, the Air Force is replacing its aircraft at a much slower rate than it did in the 1970s and the 1980s; thus, many old aircraft remain in the force (Pyles, 2003).

Table 1.1 lists some of the fleets that are the backbone of the force. Many aircraft currently in operation have been in service for more than 30 years with no retirement plans. Additionally, current plans indicate that many fleets will continue to be in service for many years. For instance, the B-52 fleet is expected to remain in operation for more than 80 years (Hebert, 2003). Although replacements have been identified for some of these fleets, a significant portion of them will not be replaced for many years to come because they are so large.

Fleet-Management Challenges with Aging Fleets

Sustaining the aircraft structure is one of the main challenges in extending use of old aircraft.[1] As aircraft age and as their use

---

[1] An aircraft's structure, its *airframe*, includes "the fuselage, wing, empennage, landing gear, control systems, engine section, nacelles, air induction, weapon mounts, engine mounts, structural operating mechanisms, and other components as described in the contract specifications," per U.S. Air Force standard for its Aircraft Structural Integrity Program (ASIP), MIL-STD-1530B.

increases, damage accumulates in their structures and various subsystems. Various damage mechanisms, such as fatigue cracking, corrosion, and stress-corrosion cracking, contribute to the deterioration of aircraft structure.[2] Other nonstructural systems, such as the wiring system, also age as materials deteriorate from environmental exposure and mechanical stresses. Many older aircraft are facing these aging issues, and many more aircraft are expected to encounter them as the Air Force keeps them in service for many years.

Structural deterioration can lead to serious problems. Structural failure can have serious consequences, including loss of aircrews. There are growing concerns about a flight safety risk because of the

Table 1.1
A Sample of Aging Fleets in the U.S. Air Force

Aircraft Type	Fleet Size	Initial Operational Capability	Average Age	Future Plans (as of 2002)
A-10	359	1977	22	Retain at least till 2028 Replace with F-35
B-52H	94	1962	42	Retain at least till 2040 No replacement identified
C-5	126	1969 (C-5A)	26	Retire in 2040 No replacement identified
C-130	679	1961 (C-130E) 1973 (C-130H)	25	Retain at least till 2030 Purchase 150 new C-130Js
KC-135	576	1957	42	Retire in 2040 No replacement identified
F-15	734	1975 (F-15A/B) 1979 (F-15C/D) 1989 (F-15E)	19	Retire by 2012 (except F-15Es) Replace with F-22
F-16	1,361	1980	14	No retirement date yet Replace with F-35
T-38	489	1961	37	Retain at least till 2020 No replacement identified

SOURCES: Air Force Association (2004), AATT (2002).

---

[2] These damage mechanisms are primarily those related to the aging of metallic aircraft structures. See Tiffany et al. (1997) for details on these mechanisms. Fatigue cracking is discussed in Chapter Two.

increasing number and extent of structural deteriorations as aircraft age and with increasing use (Pyles, 2003; Hebert, 2003). There are further concerns about increasing demand for inspections, repair, and modifications. Such increases in maintenance workload may become extremely costly and may lead to poor aircraft availability, threatening mission capabilities (Hebert, 2003; Jumper, 2003; Secretary of the Air Force, 2002). In the past, the Air Force has imposed flight restrictions and grounded aircraft for the sake of aircraft safety after serious fatigue damage has been found (Pyles, 2003).

The Air Force thus faces the fleet-management challenge of converting these increasing structural needs to resource requirements, such as maintenance, modification, and replacement plans, to prevent structural failure, future economic burden, and loss of mission capability.

The U.S. Air Force Aircraft Structural Integrity Program

A series of catastrophic wing structural failures on B-47s (see the appendix) drove the Air Force to establish its ASIP in 1958.[3] ASIP's overarching objective is to prevent loss of operational effectiveness and minimize costs while maintaining structural safety (i.e., preventing catastrophic structural failure) (AFPD 63-10).

ASIP provides a framework for ensuring the structural integrity of an aircraft throughout its life, from acquisition to retirement, and outlines the basic principles of structural-life management. During the acquisition phase, ASIP activities involve design, analysis, and tests to ensure that the aircraft structure is adequate to operate as intended. During the sustainment phase, ASIP activities involve the data collection, analysis, and tests needed to plan such continuous sustainment activities as maintenance and modifications to ensure that the structure remains safe until retirement. These activities pro-

---

[3] Note that the program name and its acronym are also used both for similar programs (e.g., that of the Canadian Forces) and for structural integrity programs tailored for specific aircraft types. The actual name of a tailored program would include the type designator.

vide information about structural conditions that can aid fleet-management decisions, such as creating inspection and maintenance plans, setting modification priorities, and enabling the decisionmakers to consider structural safety, cost, and operational effectiveness properly in managing the fleet.

The Air Force has been successful at ensuring structural safety for decades through ASIP. ASIP is a key contributor to the force-management processes, and the program's viability will be critical as the Air Force continues to operate an aging force to meet the nation's needs.

Current ASIP Challenges

In recent years, concern has arisen about ASIP's ability to continue meeting the Air Force's needs, given the aging force, budget pressures, and diminishing control of the servicewide ASIP (Tiffany et al., 1997). These factors are interdependent and together strongly influence the program's effectiveness.

Because the Air Force plans to fly many of its aircraft for many years, there is an increasing demand for more accurate knowledge about the current and future structural conditions of individual aircraft and the associated risks of structural failure. The need for engineering capabilities, both in terms of research and development and in terms of engineering analysis, is increasing as age-related problems grow. A 1997 National Research Council (NRC) study on the Air Force's aging force (Tiffany et al., 1997), as well as the engineering community at recent ASIP conferences, has voiced a concern that budget pressures, rather than structural needs, are driving the level of ASIP implementation because fleet managers must allocate available resources among multiple elements, such as sustainment of aging airframes and other aging aircraft subsystems (e.g., modernization of avionics).

Furthermore, control of ASIP has been diminishing over the years as a result of reforms in the 1990s to minimize government acquisition regulations. Prior to these acquisition reforms, Air Force

Regulation (AFR) 80-13 and the ASIP standard, MIL-STD-1530, were used to enforce ASIP. The regulation ensured compliance with ASIP within the Air Force. The military standard required contractor compliance because it could be included in the contracts. However, with the acquisition reform, the AFR was converted to an Air Force Instruction (AFI) and the ASIP military standard (MIL-STD-1530) to a military handbook (MIL-HDBK-1530B). That change meant it could no longer be cited as a contractual requirement. As a result, the industry and the contractors, as well as the system program offices that carry out the ASIP, interpreted the former requirements as guidelines.

At the same time, other ASIP oversight activities have diminished. There have been concerns that the reduced authority and lack of oversight could lead to undesirable variability in ASIP implementation (omitted or incomplete ASIP tasks), especially in light of budget pressures.

These concerns about ASIP are compounded by the challenges the engineering community currently faces in communicating with decisionmakers. At recent ASIP conferences, the engineering community has expressed that one of the main challenges in structural-life management processes has been communicating structural integrity issues to decisionmakers. Community members cited several potential causes, including the following (ASIP Conference, 2003):

- lack of technical understanding by the decisionmakers
- insufficient data on structural conditions
- lack of resources to gather sufficient information on structural conditions
- lack of an outlet for communicating key structural integrity issues to the decisionmakers.

As a result, decisionmakers may not fully understand the fleets' structural conditions or the consequences of inadequate ASIP implementation.

The decisionmakers, on the other hand, may not be concerned about ASIP's effectiveness, since there have been no catastrophic structural failures in recent years. However, the concern is that the reduction in ASIP regulation, communication challenges, and budget pressures may lead to loss of control of the program as a whole, lack of awareness of structural conditions, and resource allocation challenges for those managing structural life. As a result, ASIP may not be implemented adequately (e.g., omission of or incomplete ASIP tasks), degrading its effectiveness.

These challenges, and the Air Force's need to continue to operate an aging force for many years to come, raise concerns about the future effectiveness of ASIP and the effects that may have on the force's operational effectiveness, aircraft safety, and sustainment costs.

The Air Force has begun to address some of these challenges. For example, in February 2004, the Aeronautical Systems Center's Engineering Directorate (ASC/EN) converted MIL-HDBK-1530B back to a military standard so that it could once again be used as a contractual requirement.[4] This will reestablish some standardization of and control over ASIP. Further, the Air Force is in the process of institutionalizing a framework for structural risk assessment to enable communication of structural risks to decisionmakers, which will help them better understand structural integrity issues.[5]

Study Purpose and Scope

The purpose of this research is to provide insights and guidance on how the Air Force can continue to strengthen and improve ASIP to meet the program's objectives in the presence of current challenges and needs.

Although some structural-life management issues that arise during the sustainment phase may be due to choices made during the acquisition phase (e.g., selection of material for performance rather than durability), this report focuses primarily on the sustainment

---

[4] The Air Force released MIL-STD-1530C in November 2005, as this report was being prepared for publication. These changes do not significantly affect our findings.

[5] The new version, MIL-STD-1530C, now requires ASIP to include a risk analysis and updates and that the results be reported for formal acceptance.

phase to address current ASIP challenges for the aging force. The research scope is thus limited to aircraft that are no longer being procured.[6]

The technical scope of ASIP is expanding to address other issues related to structural integrity, such as composite materials and corrosion. This report, however, focuses on ASIP processes that deal with metal fatigue.

Postulating that the regulations, communications between the structural-life management authorities, and the level of resources available influence the effectiveness of ASIP, we surveyed aircraft structural-life management programs in other military services to broadly understand their approaches to structural-life management, focusing on regulation, communication, and resource management.

The effects of these approaches on ASIP effectiveness are difficult and complex to quantify because of the large scope of the program and the effects of multiple variables, such as operational factors, technical scope, and the multiple elements involved in fleet management (e.g., depot organizations). This research is limited to a qualitative analysis of the implications of the different approaches to gain insights on how the Air Force can strengthen ASIP.

Approach

We believe that three interdependent factors promote adequate and effective ASIP implementation: (1) independent and balanced regulation, (2) clear and timely communications, and (3) adequate and stable resources (Figure 1.1). We therefore sought insights from other military aircraft fleets by identifying the mechanisms or approaches they use in these areas.

---

[6] The research also does not include the modified commercial aircraft in the U.S. Air Force and U.S. Navy inventories. These were designed to meet Federal Aviation Administration (FAA) requirements and thus may follow FAA maintenance requirements and other FAA regulations that differ from the services' own structural-life management programs.

Figure 1.1
Critical Program Management Elements to Achieve ASIP
Effectiveness

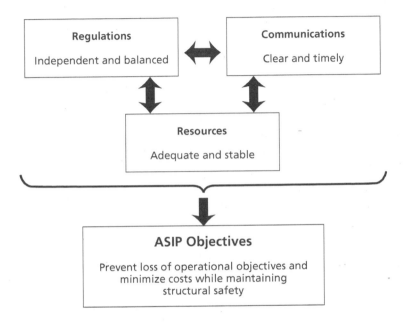

Selection Method and Data Collection

We selected the structural-life management programs in the U.S. Navy and the Canadian Forces for a comparison with that of the U.S. Air Force. The U.S. Navy was selected primarily because of its different technical approach to aircraft design and its sustainment concept. Additionally, the U.S. Navy is subject to U.S. Department of Defense (DoD) policies and budget pressures. We had learned that the Canadian Forces took a strikingly different approach (with a strong regulatory aspect) to structural-life management. Researching these services was convenient for data collection and provided a broad enough perspective on a variety of approaches for research purposes.

We collected information on the processes, organizational structure, policies, and technical methods of the three aircraft structural-life management programs. This information came from site visits, discussions at ASIP conferences, the literature, and service publications on policies and procedures. We did not collect any financial data and thus did not conduct cost analyses for comparison.

Potential Bias

The limited scope of the presented research focused our attention on factors that have been raised as key issues for the Air Force in recent years. We specifically sought ways to address those issues. As a result, there may appear to be a bias against the Air Force's approach. However, this research does not attempt to identify the relative effectiveness of the programs compared. We sought to minimize the potential for bias by identifying the potential positive and negative implications of all three approaches. Additionally, we tried to gain a full understanding of the potential effectiveness of each approach by assessing it within the contexts of the specific program and the service's operations.

Report Organization

Chapter Two begins by describing the technical basis for aircraft structural-life management programs, which affects various aspects of the program, then outlines the basic elements of such a program. The next three chapters describe each service's structural-life management program in turn: the U.S. Navy's program in Chapter Three, the Canadian Forces' program in Chapter Four, and the U.S. Air Force's program in Chapter Five. Each discussion includes operational factors; the program's structure and principal elements; and an explanation of how regulations, communications, and resource management come into play. Chapter Six makes some observations about the different approaches, summarizing each. Chapter Seven presents possible actions the U.S. Air Force could take to strengthen its ASIP.

CHAPTER TWO

Background

This chapter begins with a background of aircraft structural design concepts, specifically those related to fatigue, and how they affect structural-life management. We then describe the basic elements of an aircraft structural-life management process. Different fatigue design concepts lead to different approaches to the elements of aircraft structural-life management, such as inspection and maintenance plans.[1]

Fatigue as a Limiting Structural-Life Factor

Fatigue is one of the primary mechanisms causing deterioration of an aircraft's structure during its lifetime. Landings, takeoffs, and maneuvers subject the structure of an aircraft to repeated stress over its lifetime. This is referred to as cyclic or alternating loading (Figure 2.1). Cyclic loading causes fatigue cracks to form in the structure. These cracks grow longer with each stress cycle, degrading the aircraft's structural strength. A crack initially grows slowly, but the rate accelerates as these cycles accumulate, to the point at which rapid crack growth results in a fracture (i.e., the part breaks). Thus, one of the key design criteria for an aircraft is that it can endure accumulated fatigue damage over its service life to prevent structural failure.

---

[1] Readers who are already familiar with this subject may choose to proceed to the next chapter. Those who require a more in-depth discussion should refer to such works as Broek (1986), Broek (1988); Gallagher et al. (1984); Dowling (1993); and Grandt (2003).

Figure 2.1
Fatigue Crack Grows Under Cyclic Loading, Leading to a Structural Failure

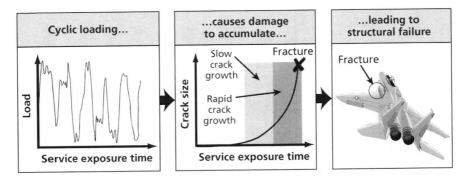

Fatigue-Design Concepts: Damage Tolerance Versus Safe-Life

There are two distinct design approaches for protecting an aircraft's structure from failure due to fatigue damage: safe life and damage tolerance (Figure 2.2). These fatigue design approaches differ in how they model damage growth, what they assume about the initial conditions of materials, and the failure criteria they use to establish the aircraft's original design service life.

The fundamental difference between these is that the safe-life approach assumes that no fatigue cracks will exist in the structure during the specified lifetime for safe operation; the damage-tolerance approach assumes that potential fatigue cracks may exist in critical locations in fracture-critical parts. For practical purposes, the safe-life approach does assume that very small fatigue cracks may exist. The damage-tolerance approach requires that the structure be able to tolerate slowly developing cracks safely, until they can be detected and repaired.

Figure 2.2
Comparison of Safe-Life and Damage-Tolerance Fatigue Approaches

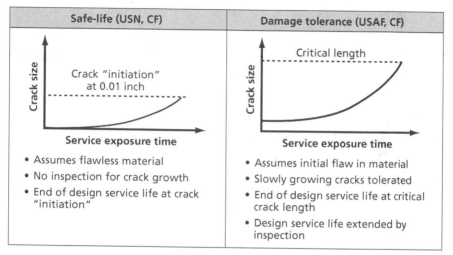

NOTE: Many Canadian Forces aircraft are derived from those of the U.S. Navy. Their overall approach is a hybrid, as discussed in Chapter Four.

Safe-Life: The Choice of the U.S. Navy and the Canadian Forces

The safe-life approach to fatigue design requires analysis or testing to demonstrate that the probability of failure is remote and that no detectable cracks will exist during the specified structural lifetime. Such a general requirement allows the failure criterion, load spectrum,[2] fatigue analysis method, and life-reduction factor used in establishing the design life to vary.

This subsection describes the U.S. Navy's safe-life approach (Figure 2.2).[3] This approach assumes that the material is initially flawless and uses the initiation of a crack as the failure criterion. The Navy defines a crack as having been initiated when it is 0.01 inch long. The mean time for a crack to reach this length is determined

---

[2] The *load spectrum* is defined as a sequence of loads with specific ranges.

[3] See Rusk and Hoffman (2001) on the U.S. Navy's safe-life approach. Note that the Canadian Forces use a similar approach because many of their aircraft are derived from those of the U.S. Navy.

from full-scale fatigue tests, in which expected service loads are simulated and applied to an aircraft in a laboratory environment. This test-demonstrated fatigue life is divided by a life reduction factor of two to arrive at the design service life.[4] The life-reduction factor accounts for variability in both material properties and fatigue loads. The Air Force used a safe-life approach in the 1950s, using *fracture* as the failure criterion (defining it as the point at which a critical part breaks) and a life-reduction factor of four.

To further account for the variability in the loads that an aircraft undergoes in service, the U.S. Navy uses a "severe" load spectrum rather than an "expected" load spectrum to establish the design service life, even though only a small percentage of the fleet is exposed to severe loads in service. The load spectrum is based on historical data from a similar fleet that has flown similar missions, with adjustments for relative severity.

The U.S. Navy uses the local strain method to calculate the accumulation of fatigue damage.[5] The approach the U.S. Navy and the Canadian Forces take to calculating safe life is therefore sometimes referred to as the *strain-life approach.*

Theoretically, the safe-life approach does not require safety inspections for fatigue cracks. Although the U.S. Navy inspects structures primarily for corrosion, it also, in practice, inspects some critical areas for fatigue cracks.

Damage Tolerance: The Choice of the U.S. Air Force and the Canadian Forces

The damage-tolerance approach assumes that flaws exist in the structure because of manufacturing defects and in-service activities (such as repairs) and that these flaws will cause cracks to develop during the aircraft's service life.[6] To prevent undetected flaws from causing

---

[4] Note that the Navy defines *fatigue life* as the time it takes for cracks to initiate.

[5] See Dowling (1993) for details on the local strain method.

[6] See Gallagher et al. (1984) and Air Force Research Laboratory (2002) for more details on the damage-tolerance approach.

structural failure, a damage tolerance requires that structures be able to tolerate some minimum damage for a specified period of service without being repaired, referred to as the *safe crack-growth period.*

The damage-tolerance requirement can be met with either a slow-crack-growth or a fail-safe design. In the former, the assumed initial flaw in the structure must not grow to critical size, the point at which structural failure would occur, for a specified period of unrepaired service.[7] The specified critical crack size is based on the minimum residual strength required for the structure to withstand the relatively rare occurrence of a design-limit load. In a fail-safe design, rapid crack growth must not cause complete structural failure. The surrounding structures are designed to be strong enough to withstand the loads to which they are subjected until the failed structure has been discovered and repaired.

The U.S. Air Force conducts full-scale fatigue tests to demonstrate the fatigue life (the time it takes a crack to grow to a critical size) using the expected aircraft use based on historical data. The test-demonstrated fatigue life is divided by a life-reduction factor of two to arrive at the design service life.

If a structural element can be inspected, its service life may be extended by performing inspections at intervals not greater than half the time required for a crack to grow from minimum detectable size to critical size.

Principal Elements of the Aircraft Structural-Life Management Process

Regardless of how fatigue design is approached, the structural-life management process has three basic elements: data collection, aircraft structural condition tracking, and fleet management (Figure 2.3).

---

[7] The U.S. Air Force typically assumes an initial flaw of 0.05 inch for a damaged fastener hole. The safe-crack-growth period is thus driven by the time it takes for an initial flaw to grow to a critical size (Figure 2.2).

Figure 2.3
Elements of Aircraft Structural-Life Management

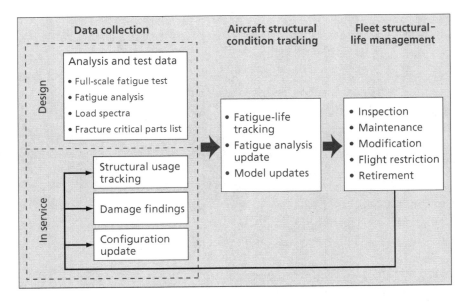

Because assumptions about expected aircraft use, the initial conditions of materials, and the damage accumulation process are made during aircraft design, tracking an aircraft while it is in service and continually assessing its structural condition throughout its service life are key to managing its structural life.

Fatigue analyses and full-scale tests are conducted in the development phase to establish the design service life and to identify fracture-critical parts for a specified load spectrum. A baseline engineering model is developed to predict how fatigue damage will accumulate over the aircraft's service life. The results from these tests are used to calibrate the engineering model and verify the initial fatigue analyses.

As shown on the left side of Figure 2.3, coupling fatigue analyses and full-scale fatigue tests during the design phase with tracking an aircraft's use, maintenance records, and configuration provide the required input data for an accurate assessment of its structural condition throughout its service life.

While the load spectrum assumed for the design process will have been based on how users expect to fly the aircraft, the actual loads the aircraft is subjected to almost always deviate from that spectrum. Thus, accurate assessment of an aircraft's structural condition over time requires continuous collection of structural data while the aircraft is in service. These data may be captured by instrumentation on the aircraft structure. Some aircraft are even equipped with strain gauges for major components, such as the wings and vertical and horizontal stabilizers. Flight load measurements can be used to derive an actual load spectrum for the structure. It is also possible to approximate a load spectrum using data on flying hours, types of maneuvers, and numbers of landings and takeoffs, although this method is not as accurate as using actual flight load data.

After an aircraft enters service, its configuration is likely to change over time (e.g., modifying the structure to accommodate additional payloads), causing deviation from the tested configuration. If the current configuration has not been modeled accurately, the predicted fatigue damage may be grossly miscalculated. Thus, information on configuration changes must be collected to update the engineering model. Data from maintenance and inspection are also collected to provide information on any unanticipated structural damage and to verify the predicted damage accumulation.

As the middle of Figure 2.3 shows, an engineering model uses test and structural-use data to estimate how the structure accumulates fatigue damage. Because the engineering models are calibrated on the initial tests, there may be some associated uncertainties. If the actual structural use differs greatly from the loads initially assumed (the design loads), the uncertainties about the structural condition may be significant. Periodic assessments of the engineering models and the structural condition, including new, full-scale fatigue tests or extensive inspections, may be necessary to verify the condition of the structure and to ensure that the engineering models continue to be reliable.

As the right section of Figure 2.3 shows, fleet-management plans are then developed using the information on current and predicted structural conditions. Maintenance plans and inspection intervals are determined for the critical locations using the fatigue damage accumulation rate. These plans ensure safety and minimize any unscheduled maintenance that may lead to costly repairs. Decisions regarding flight restriction are also based on the structural assessment. Thus, structural condition tracking helps decisionmakers allocate resources for sustaining structural integrity appropriately, such as in scheduling modifications and phasing individual aircraft into and phasing out of a fleet.

All three services are currently moving toward incorporating some form of risk assessment into their structural-life management processes to assist fleet-management decisions. The U.S. Navy is researching application of probabilistic methods for quantifying aircraft structural risks (Rusk and Hoffman, 2001). The Canadian F-18 program office currently uses structural risk assessment to assist its fleet-management decisions. The Air Force is in the process of institutionalizing a risk-based framework into ASIP using a probabilistic approach as a means of assessing structural risks.[8]

The cycle of in-service data collection, structural-life tracking, and fleet-management planning processes repeats throughout an aircraft's service life. The structural-life management plans are continually updated as more data become available from ongoing monitoring of the structural condition.

---

[8] A probabilistic approach to damage tolerance uses statistics to account for nature of crack growth, material properties, and initial crack distribution in the material to determine the probability of failure. MIL-STD-1530C now requires the ASIP process to include risk assessment.

CHAPTER THREE

Aircraft Structural-Life Management in the U.S. Navy

The Navy operates approximately 2,000 fixed-wing aircraft of about 20 different aircraft types, many based on carriers (U.S. Navy, 2005). The Navy has taken the safe-life approach to determining the structural-life limits and inspection policies for its aircraft. Because space for equipment, supply, and maintenance personnel is limited aboard aircraft carriers, the inspection and maintenance environment is austere. Because the safe-life approach requires minimal routine inspection for fatigue cracks, it has the advantage of offering periods of maintenance-free operation without compromising safety.

This chapter describes the U.S. Navy's approach to aircraft structural-life management. We examine the governing policy, the authorities, and the key processes to understand the characteristics of regulations, communications, and resource-management approaches in the U.S. Navy's aircraft structural-life management.

Program Structure

A Naval Air Systems Command (NAVAIR) instruction outlines the policy, rules, and procedures for establishing and maintaining the structural integrity of all Navy aircraft (NAVAIRINST 13120.1). The instruction describes the principal elements of structural-life management and assigns responsibilities to various organizations. A centralized program, Aircraft Structural Life Surveillance (ASLS), carries out the majority of these tasks for all Navy aircraft.

Governing Policy

The governing policy behind the Navy's approach to structural-life management is that the aircraft must not exceed the structural-life limits during service to ensure structural safety. The Office of the Chief of Naval Operations (OPNAV) instruction (OPNAVINST) establishes this policy (OPNAVINST 3110.11).

NAVAIR establishes the structural-life limits in terms of flight hours and catapult and arrestment use for the airframe and various fracture-critical components for each aircraft type (more specifically, for each "type, model, and series" [T/M/S]).[1] These limits are based on the fatigue-life limits for the airframe and critical components. The Navy then manages the structural life by tracking aircraft use and calculating the accumulated fatigue damage for every aircraft by tail number, resulting in a quantifiable metric, the fatigue-life expended (FLE). The FLE is the complement of fatigue-life remaining, that is, an FLE of 100 percent is equal to 0 percent fatigue-life remaining.

Once an aircraft reaches its structural-life limit (100-percent FLE), the Navy chooses either to retire the aircraft or to extend its structural life through modifications. Inspection is not an option for extending structural life under the safe-life philosophy.

Principal Authorities

Figure 3.1 is an organizational diagram, with key aircraft structural-life management authorities highlighted. The dotted line indicates a supporting relationship, while a solid line indicates a chain of command.

Type commands (TYCOMs) are the key commands within the Navy responsible for managing aircraft operations on each coast.[2] The TYCOMs are highly involved with logistical and maintenance requirements (including budget and schedule management for depot-level maintenance).

---

[1] This is not to be confused with the term *type model series* (TMS), used to designate aircraft engines.

[2] These are specifically the Commander U.S. Naval Air Forces Atlantic (AIRLANT) and the Commander U.S. Naval Air Forces Pacific (AIRPAC).

Figure 3.1
Principal Organizations in the U.S. Navy's Aircraft Structural-Life Management

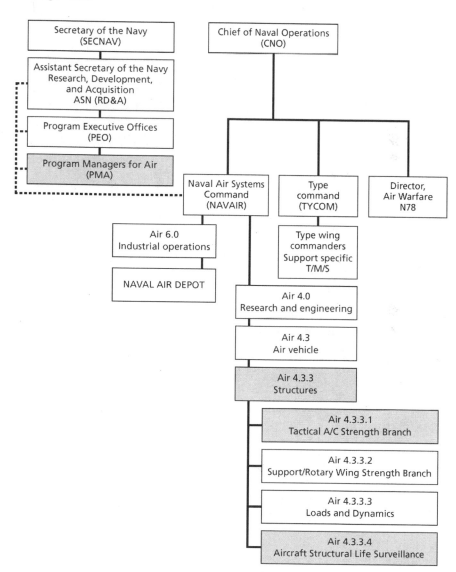

Each T/M/S fleet has a program manager for air (PMA), who heads the office responsible for that fleet's total life-cycle management. The PMA has the ownership and decision authority for the fleet's structural-life management (except for fatigue-life tracking, which is the responsibility of Air 4.3.3.4) and must comply with the policies outlined in NAVAIRINST 13120.1. The PMA works closely with programmers and requirements officers in the Office of CNO to project adequate annual funding for supporting fleet requirements.

NAVAIR Air Vehicle Department (Air 4.3) supports PMA management of aircraft structural life, with the responsibilities distributed among several organizations within Air 4.3. The Structures Division (Air 4.3.3) carries out a majority of the structural-life management tasks and has regulatory responsibility for the technical aspects of structural-life management, which will be discussed later in this chapter. The ASLS program (Air 4.3.3.4) carries out fatigue-life tracking for all Navy aircraft. A structural lead engineer from the Strength Branch (Air 4.3.3.1) is assigned to the program office to support each PMA's aircraft structural-life management activities. The structural lead coordinates with the PMAs and the branches in the Structures Division, such as ASLS, to support structural integrity.

All Air 4.3 engineers work in a single facility at Patuxent River, Maryland. The only engineering staff stationed elsewhere are the fleet support team (FST) engineers, who are located at the various depots. The FST engineers have the authority to approve local engineering work (e.g., one-of-a-kind field repairs) but communicate fleetwide issues affecting a given T/M/S to the Structures Division at Patuxent River, maintaining clear organizational lines. The PMAs also work in NAVAIR facilities but are not part of the NAVAIR chain of command. They maintain their statutory lines of authority reporting to the program executive officers (PEOs).

Aircraft Structural-Life Management Process

Figure 3.2 illustrates the principal elements in the structural-life management of naval aircraft:

Figure 3.2
The U.S. Navy Aircraft Structural-Life Management Approach

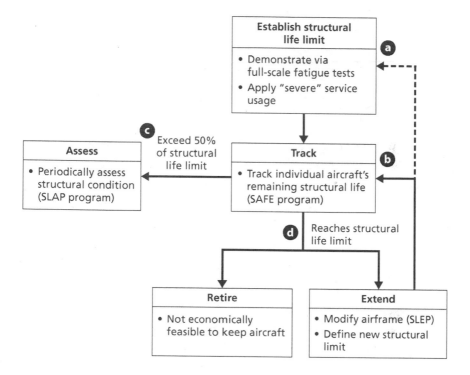

- establishing structural-life limits
- tracking structural life
- assessing structural life
- retiring aircraft or extending their service lives.

The ASLS program carries out the last three tasks for all Navy aircraft. The ASLS program (1) continuously tracks the fatigue lives of individual aircraft under the Structural Assessment of Fatigue Effects (SAFE) program, (2) periodically reassesses aircraft structural conditions and the tracking program under a Structural Life Assessment Program (SLAP), and (3) provides structural integrity support for the Service-Life Extension Program (SLEP).

Establishing Structural-Life Limit

As noted in Chapter Two, the U.S. Navy bases its structural-life limits on a severe spectrum of design loads. The Navy uses test data on fatigue life to determine the flight-hour limits for various fracture-critical components and the catapult and arrestment limits for each T/M/S. These limits are then published in NAVAIRINST 13120.1.

Structural-Life Tracking

The Navy uses FLE to track accumulated fatigue damage to monitor an aircraft's structural condition. The SAFE program tracks FLE continuously for all aircraft by tail number. The funding NAVAIR Air Vehicle Department (Air 4.3) receives for this program is separate from that for the PMA budget-planning process for the fleet. As a result, this program can independently assess fatigue life for all aircraft.[3] Figure 3.3 provides an overview of the SAFE process.

The fatigue-life limits for individual aircraft vary from those established for their respective T/M/S because how individual aircraft are actually used in service will vary from the design assumptions. To account for the variation, the Navy tracks the structural use of each aircraft and analytically computes its FLE. Flight loads, landing, catapult, and arrestment data are collected for every aircraft. NAVAIR Logistics maintains the data for all fleets. Operating squadrons are

Figure 3.3
SAFE Program Process

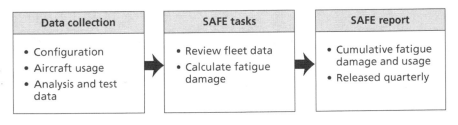

---

[3] The separate funding also lends stability to the activity.

responsible for reporting these use data for their aircraft. NAVAIR-INST 13920.1 sets formal procedures for collecting and submitting flight-load data.

SAFE also requires both configuration data and applicable engineering data from the full-scale fatigue tests. The lead structural engineer supporting the PMA on structural integrity provides information on configuration changes, aircraft use changes, and maintenance findings to ASLS to ensure that the SAFE analysis properly reflects these data.

The SAFE program gathers all flight-load data and reviews them for quality control. These data are then used to document aircraft use histories and to calculate the FLEs for all Navy aircraft. When the required flight-load data are missing for some aircraft (e.g., incomplete data were submitted), SAFE accounts for the data by assuming worst-case values in the FLE calculations for the aircraft. Thus, incomplete data may lead to an assumption that fatigue life is being expended more quickly than is actually the case. To minimize such situations, the PMA communicates with the operators to ensure timely submission of accurate and complete data.

The SAFE report formally documents FLE information for all Navy aircraft. The report is distributed quarterly to a wide range of Navy organizations, including operators, PMAs, and maintainers. The formal distribution of information ensures timely communication of structural conditions to the appropriate authorities. In some instances, a PMA may request more-frequent FLE updates than the regular SAFE report offers.

The SAFE report is a key aid for structural-life decisionmakers. It profiles FLE distribution in the fleet and thus helps prioritize modifications and phasing individual aircraft into and out of a fleet.

Structural-Life Assessment

To ensure that fatigue life is being tracked accurately, the Navy requires periodic assessment of a fleet's overall structural condition. The PMA is responsible for conducting a SLAP to evaluate the current structural condition of the fleet and to verify its remaining fatigue life. A SLAP may also be required if the mission or the use of the aircraft has significantly changed from the original design.

ASLS develops an assessment plan and carries it out for the PMA. The program may involve a wide range of activities, such as an assessment of in-service use, a teardown inspection, laboratory tests, and an analysis update. In some cases, ASLS also recommends conducting a full-scale fatigue test. Typically, the in-service aircraft that has seen the most severe use, in terms of fatigue damage, is selected as the test article. To verify the condition of the test aircraft's structure, a teardown inspection and test data evaluation follow the fatigue test. A SLAP can be a multiyear effort, especially if a full-scale fatigue test is involved. For example, the P-3C SLAP spanned five years (Shah, 2004).

ASLS then uses the SLAP results to define what needs to be done to ensure structural integrity, with respect either to the current structural-life limit or to any potential life extension (via SLEP). The SLAP results are also used to revise and refine the SAFE program's models and analyses and to revise the aircraft maintenance program, if necessary.

In the past, SLAP results have shown fatigue cracks occurring earlier than predicted. As a result, ASLS recommends a SLAP when the majority of a fleet has reached 50-percent FLE to provide a sufficient lead time for a SLEP, if needed. This minimizes the risk of aircraft reaching its structural-life limits before a SLEP can be completed, thus avoiding flight restrictions or grounding. The PMA uses a trend analysis of the fleetwide average FLE to estimate when a SLAP and a SLEP would be necessary.

A SLAP primarily benefits a particular fleet, verifying its overall structural condition and providing the PMA sufficient information for sustainment planning. However, because ASLS centrally manages the SLAPs for all the fleets, applicable information from one SLAP is shared with other fleets. For example, some of the fleets may share similar design features, or a particular fleet's SLAP results may lead to an improvement in the fatigue analysis method. This would benefit all the fleets because the centralized program uses the same methodology.

Service-Life Extension or Retirement

The Service-Life Extension Process. Once an aircraft reaches its structural-life limit, the PMA weighs the operational requirements to evaluate whether it is more economical to perform a SLEP or to retire the aircraft. The PMA matches projected fleet demands with projections of remaining service life to determine the structural needs that would meet these requirements. The PMA then funds a SLAP and a SLEP if these are required to meet the structural needs.

A SLEP requires modifications or replacements of critical components. Additionally, depending on the extent of the modifications or replacements, a new full-scale fatigue test may be conducted to establish the extended structural-life limit.

Using the SLAP results, ASLS develops modification requirements and plans, working with the PMA. Ultimately, NAVAIR Structures must certify the modifications to ensure that the aircraft can maintain structural integrity to the extended service-life limit. This certification provides an independent technical assessment of the PMA's decision.

The PMA prioritizes and schedules modifications for various critical components according to the FLE of the individual aircraft. Individual FLEs also allow the structural conditions of the aircraft within a fleet to be compared.

Aircraft Retirement Process. OPNAVINST 5442.8 describes the principal organizations involved in the management of active and reserve inventories, including those involved in retirement decisions. The TYCOMs, working closely with OPNAV staff aircraft-inventory managers and PMAs, make recommendations for the disposition of aircraft approaching the ends of their service lives (OPNAV 5442.8, p. 9). "Strike boards" take place twice a year; these include representatives from the TYCOMs, PMAs, OPNAV inventory managers, and other involved agencies. The boards select aircraft to be removed from the active inventory. The costs associated with sustaining the aircraft and its operational effectiveness often influence the TYCOM's decision. OPNAV staff process the board's recommendations and send them to the Director of Air Warfare (OPNAV N78) for approval before the CNO officially removes the aircraft from service. The PMA then uses FLE information to phase out the aircraft.

Risk Management

The U.S. Navy does not have a formal periodic review process for assessing structural risk. In the past, the strict policy on structural-life limits has precluded the need for assessing structural risk beyond the end of fatigue life. With the increasing need to extend the service lives of many aircraft, a need for understanding risks beyond the aircraft's fatigue-life limit emerged. NAVAIR is currently in an early stage of research on applying probabilistic methods for quantifying aircraft structural risks.

The Navy's risk-management approach is outlined in NAVAIRINST 5000.21. This framework is similar to that of the U.S. Air Force (see NAVAIRINST 5000.21). Structural risk is quantified in terms of the probability of catastrophic structural failure on any given flight. Before the end of the aircraft's fatigue life, the probability of catastrophic structural failure is assumed to be less than 1 in 10 million,[4] which is the acceptable structural risk level NAVAIR has established.

Use of a distributed authority structure means that risk is managed at various levels. The NAVAIR Air Vehicle Department provides all the requisite data and risk assessment predictions to the PMAs and PEOs and the Commander, NAVAIR via Research and Engineering (Air 4.3/4.0) chain of command. The appropriate authority uses these predictions to decide whether to accept the risk or whether to set out a risk-mitigation plan, such as flight restriction and modifications.

This process was used during Operation Iraqi Freedom, when EA-6B's were flown at elevated risk levels with approval from the Commander, Naval Air Forces and CNO. NAVAIR used a simplified

---

[4] This probability of failure is based on a 1 in 1,000 probability of failure of a single specimen within its lifetime.

probabilistic method to assess the structural risk of operating some EA-6Bs beyond the ends of their fatigue lives. Once combat operations were over, the CNO mandated that the acceptable risk levels be reset to those NAVAIR had established, thus grounding aircraft exceeding these limits to reduce the risk until repair or replacement of structural components.

Regulatory Processes

In the U.S. Navy, oversight of structural-life management activities is minimal, as is assessment of compliance, and works through the existing chain of command. For example, the head of NAVAIR's Structures Division (Air 4.3) oversees the ASLS program (Air 4.3.3.4).

The Structures Division is responsible for overseeing the technical aspects of structural-life management. The division is the final authority on structural integrity and must certify any structural changes to ensure that integrity is maintained until the established structural-life limit has been reached. For instance, an aircraft must be recertified through a SLEP if it is to operate beyond its FLE limit. Recertification is also required for structural modifications. The Structures Division determines the criteria for recertification on a case-by-case basis (e.g., structural analysis, component testing, or full-scale fatigue tests).

Because the Structures Division also provides structural integrity support to the PMAs, assigning oversight authority to the division creates a potential conflict of interest. However, this is mitigated by the fact that the division and the PMAs are in different chains of command. Additionally, a centralized program, ASLS, carries out a majority of the structural integrity support using largely standardized processes and procedures for all the aircraft fleets. These arrangements promote maintaining independence from the PMAs in the certification process.

Summary

The characteristics of the U.S. Navy's approaches to aircraft structural-life management can be summarized as follows:

- Regulation
  - The Navy's policy on structural-life management is explicit.
  - The visibility of FLE data ensures compliance.
  - A central authority governs the technical aspects of structural-life management.
- Communication
  - A single, standard metric, FLE, is used to convey structural conditions.
  - Fatigue life is tracked rigorously, and the resulting data are disseminated widely through frequent formal reports.
  - Close working relationships and colocation promote and facilitate informal communication.
- Resource management
  - The structural-life monitoring has dedicated funding.
  - FLE data are used to help prioritize resources.
  - The policies that drive key resource-allocation decisions are explicit.
  - Independent technical assessments aid resource allocation.

Regulations

The Navy has made its policy on maintaining the structural integrity of all aircraft by defining a quantifiable metric describing the structural condition of an aircraft (FLE) and by setting a threshold for this state (FLE of 100 percent).

The Navy's oversight of structural-life management, however, is informal. Compliance monitoring is achieved primarily through visibility; the FLEs of individual aircraft are disseminated to all the organizations involved in aircraft operation and support. Making all FLE data available to all involved in aircraft operation and support, including senior leadership, allows these organizations to be continuously aware of the state of each aircraft.

A single centralized organization, NAVAIR's Structures Division, holds regulatory responsibility for structural integrity, providing checks and balances in structural-life management.

Communications

The Navy uses a quantifiable metric, FLE, as the primary means of communicating structural condition to those operating and supporting aircraft. Rigorous and accurate monitoring of accumulated damage is critical to the Navy because it uses the safe-life approach. In this approach, inspecting for cracks is not its primary method of validating structural conditions.

The Navy has a centralized program that independently tracks fatigue life for all aircraft and disseminates this information regularly across a wide range of Navy organizations via a formal report.

The communications between the NAVAIR Structures Division and the PMAs are primarily informal and are facilitated by the working relationships among and colocation of the principal authorities in structural-life management. The colocation of all Structures Division personnel and ASLS program staff promotes information-sharing and cross-fertilization across the different program offices.

Resource Management

Tracking the fatigue lives of individual aircraft is critical not only for providing visibility into the structural conditions of all Navy aircraft but also for providing information for key structural-life management decisions. To ensure that such critical information is continuously available throughout the aircraft's service life, the Navy funds fatigue-life tracking separately. This separate funding further enables the SAFE program to maintain its independence from PMA's assessments.

The explicit policy on structural-life limits ensures that the PMA understands the potential consequences of not planning or of untimely planning of critical tasks, such as a SLAP and a SLEP. Lack of timely planning could result in imposition of flight restrictions on aircraft or grounding them.

The PMAs use the FLE information when making resource-allocation decisions, such as determining when to conduct SLAPs and SLEPs, planning the force structure, and scheduling modifications. SLAP results also guide PMA decisions on resource allocation.

The NAVAIR Structures Division's role in the certification of structural integrity provides an independent technical assessment of the PMA's resource-allocation decisions, establishing checks and balances for the resource-management process.

CHAPTER FOUR

Aircraft Structural-Life Management in the Canadian Forces

The Canadian Forces operate about 350 fixed-wing aircraft and about a dozen different aircraft types in a land-based environment (Canada Department of National Defence, 2003). The Canadian Forces have acquired many U.S. Navy–derived aircraft, for which they initially adopted the safe-life approach. As the need to extend service life became apparent for many fleets, many weapon system program offices turned to the damage-tolerance approach to ensure safety beyond the original design service life (i.e., beyond crack initiation). As a result, the Canadian Forces manage most of their core fleets with a hybrid of safe-life and damage-tolerance approaches.[1]

This chapter describes how the Canadian Forces approach aircraft structural-life management. We begin with an overview of the program structure, as defined by the governing policy on aircraft structural-life management, and of the principal authorities for structural-life management. We then describe the principal elements of the Canadian ASIP process.[2] Finally, we summarize the characteristics of regulations, communications, and resource-management approaches for this program.

---

[1] Because the Canadian Forces have adopted the damage-tolerance approach, their aircraft are being modified to permit inspection of critical areas. These aircraft were not initially designed to allow access to critical areas for inspection because the safe-life approach does not require inspection for crack growth.

[2] The Canadian program has the same name as the U.S. equivalent and uses the same acronym.

Program Structure

In the late 1990s, Canada's Department of National Defence and the Canadian Forces established the Airworthiness Program to regulate the technical and operational aspects of military aviation to ensure the safety of their aircraft. The Airworthiness Program is modeled after Canada Transport, which regulates civil aviation in Canada and thus is equivalent to the U.S. FAA, but has been adapted for military aviation. Within the Airworthiness Program, the Technical Airworthiness Program focuses on regulating the relevant technical activities: design, manufacture, maintenance, and material support of aeronautical products.[3] The Technical Airworthiness Program governs the Canadian Forces' ASIP, providing independent regulations and oversight for it. Although the Technical Airworthiness Program is also responsible for all aircraft subsystems, this report focuses on the program's structural-integrity aspects.

The Canadian Forces' ASIP is decentralized. The weapon system program office for each aircraft type implements ASIP by tailoring the program to the specific weapon system being managed, following the applicable standards and rules supplied by the Technical Airworthiness Program.[4]

The Technical Airworthiness Manual was released in 2001 to provide an overview of the program and its processes, the technical airworthiness rules and standards, and the organizational responsibilities. The manual has been evolving since then to address deficiencies related to structural integrity standards as they arise.[5] When it was first released, there were gaps between it and the existing ASIP standards, primarily regarding sustainment. The manual has since been modified to address these gaps.

---

[3] The Airworthiness Program has two other components, Operational and Investigative. The latter monitors both the Technical and Operational components and investigates safety-related issues and incidents.

[4] In essence, each system has its own ASIP, derived from the overall program. Spoken of individually, each would properly be known by the system designator—as, for example, "the CF-18 ASIP."

[5] Canada Department of National Defence (2003) is a recent edition.

Governing Policy

The Canadian Forces' approach to structural-life management is based on the concept of *airworthiness*, as the Airworthiness Program defines it. The program's philosophy is that every aircraft must be "airworthy" throughout its service life. This means that the airworthiness of every aircraft type must be approved before it can enter service and must be maintained thereafter.

The Canadian Forces define airworthiness as a demonstration that a minimum acceptable level of aviation safety has been achieved. This acceptable level is based on the design, manufacture, maintenance, and material support of the aircraft requirements that have been defined for each aircraft type, as listed in its *basis of certification*.

The basis of certification includes requirements that must be met to achieve and sustain structural integrity, which are based on the expected use of the aircraft type. These are effectively ASIP requirements, similar to those provided in U.S. Air Force's ASIP standard MIL-STD-1530B (to be described later). Every aircraft type must develop a basis of certification and comply with its requirements throughout its service life to demonstrate continuing airworthiness. Note that continuing operation beyond the original service life requires developing a new basis of certification.

Principal Authorities

Figure 4.1 shows the organizations responsible for structural-life management and the chain of command. Dotted lines indicate supporting relationships, while solid lines indicate command relationships.

Each weapon system program office is responsible for supporting its fleet and implementing its ASIP in compliance with the structural integrity-related regulations the Technical Airworthiness Program establishes. The weapon system managers (WSMs) work with the Chief of Air Staff to provide sustainment support for meeting operational requirements. They are responsible for managing their aircraft fleets and are accountable for implementing ASIP for their own weapon systems and have the requisite funding and decision authority.

Figure 4.1
Principal Organizations for Aircraft Structural-Life Management in the Canadian Forces

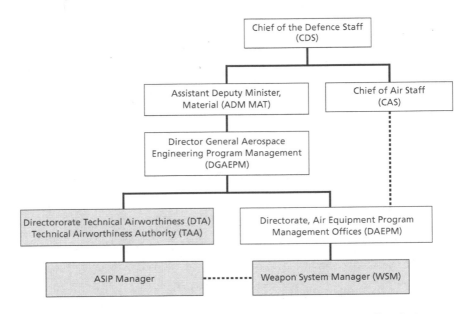

The WSMs have the flexibility to customize ASIP for their own fleets, choosing the most suitable compliance methods with the regulatory authority's approval.

Each WSM has an ASIP manager who executes the program and supports the WSM's management of structural life. ASIP managers are structures specialists in the Directorate of Technical Airworthiness, an engineering support organization in the Canadian Forces. These individuals also manage the contractors who conduct most ASIP tasks, such as data collection and analysis, structural analyses, and full-scale fatigue tests.

The WSMs' involvement in structural-life management decisions varies by WSM. The WSMs may delegate decisionmaking authority to the ASIP manager. For instance, the CF-18 (the Canadian Forces' version of the U.S. Navy's F/A-18) WSM was delegating the ASIP decision authority to the ASIP manager, while the CP-140 (the Canadian Forces' version of the U.S. Navy's P-3) program office required the WSM's approval on ASIP decisions. The WSM's responsibilities can be summarized as follows:

- customizes ASIP plans for his or her weapon system
- chooses suitable method of compliance
- authorizes funding of structural integrity-related tasks.

The Technical Airworthiness Authority (TAA) is an independent authority that manages the Technical Airworthiness Program and thus oversees the ASIPs for all the weapon system program offices. The Minister of National Defence designated the Director of Technical Airworthiness as the TAA. TAA regulatory functions include (1) establishing rules and standards for program management and technical activities affecting aircraft structural integrity, (2) assessing compliance through audits and other formal processes, and (3) accrediting organizations and individuals who conduct airworthiness-related tasks. The TAA typically consults with stakeholders in establishing rules and standards.

The Technical Airworthiness Program is not intended to restrict how organizations conduct structural-life management. The program provides general rules and standards for *what* activities need to be conducted to maintain technical airworthiness rather than for *how* the activities should be conducted. It is up to the responsible organizations to determine how to conduct these activities, and the TAA verifies on a case-by-case basis that organizations and personnel conducting technical airworthiness activities are complying with its standards.

The TAA also ensures that the personnel and organizations conducting airworthiness activities are competent and that they meet the minimum standards necessary to provide acceptable levels of flight safety through a formal accreditation process.

Because the airworthiness program is relatively new, the Canadian Forces are continually adjusting its organizations and regulations. As of March 2004, ASIP managers also served on the TAA staff. Because of the potential conflict of interest between the regulator and the implementer, the Canadian Forces were changing their organization to eliminate this dual role and to have the ASIP managers solely support their WSMs.

The ASIP managers, WSMs, and TAA are centrally located, in a single site in Ottawa, Ontario.

ASIP Process

The principal elements in the Canadian Forces' structural-life management process are as follow:

- developing the basis of certification
- monitoring structural condition
- updating the ASIP master plan
- managing risk.

Development of Basis of Certification

The establishment and sustainment of the airworthiness of an aircraft type begins with the development of its basis of certification. The WSM bases the basis of certification on the operational requirements for the aircraft, along with the applicable airworthiness standards for establishing and sustaining the structural integrity of the specific weapon system (Canada Department of National Defence, 2003). These structural-integrity requirements include the fatigue-life philosophy (e.g., safe-life), requirements for monitoring structural conditions, and acceptable risk level, among other criteria. The basis of certification must also describe requirements for adequate and appropriate maintenance and for the engineering support (e.g., data analysis) necessary for sustaining the aircraft.

Because of the Technical Airworthiness Program's inherent flexibility, the WSM and the ASIP manager develop a customized basis of certification for their specific aircraft type. Each aircraft type's basis of certification will differ from those for other types. The TAA must thus confirm the adequacy and appropriateness of each basis of certification individually.

At this point in the process, the ASIP manager also develops the ASIP master plan for the weapon system, which outlines the tasks that need to be conducted to meet ASIP requirements. These plans include data collection, maintenance, and structural analysis programs to maintain structural integrity.

Upon entering operational service, the aircraft must comply with the standards in the certification continuously. An aircraft that falls out of compliance with these standards, by, for example, exceeding its fatigue life, will require a new basis of certification if it is to continue flying. The new basis of certification may entail modifying the aircraft structure; reassessing the structure, either through full-scale fatigue tests or structural analyses; or changing the maintenance program to include inspection of critical areas. The WSMs have the flexibility to propose an appropriate means of compliance. For example, most of the CP-140 fleet is past its original (design) service goal. As of March 2004, the program office was inspecting critical areas to maintain airworthiness. A structural-life assessment program and examination of modification options were in process to develop a new basis of certification for extending the aircraft's service life.

Structural Condition Monitoring

The Technical Airworthiness Program requires monitoring in-service structural conditions. The level of this monitoring varies by weapon system, as dictated by its tailored ASIP.

The CF-18 and CP-140 weapon system offices track structural-use data for each of their aircraft. Each office has a depot contractor who maintains and analyzes the data, which the ASIP manager reviews for quality control. As in the U.S. Navy, these offices track the remaining fatigue life of critical components for every aircraft using these data. The Canadian Forces also have a metric for this purpose, which they call the fatigue-life index (FLI) or fatigue-life expended index (FLEI), which is equivalent to the U.S. Navy's FLE metric. However, unlike their peers in the United States, the Canadian Forces have not set a threshold for FLEI/FLI because of their later adoption of the damage-tolerance approach. For instance, exceeding an FLI of 100 percent does not mean that the aircraft is no longer airworthy. The actual airworthiness can be maintained through a modified inspection program that monitors components that have exceeded 100-percent FLI. Because of this lack of an FLI threshold, the Canadian Forces also use risk (in terms of probability of structural failure) as a metric to convey the state of the structural condition.

Both the CP-140 and CF-18 program offices document the FLI/FLEI in their quarterly ASIP reports. Structural-use tracking has enabled the CF-18 program office to provide feedback to operators on aircraft use to minimize the damage on the structure. The program office, for example, may identify certain maneuvers that are not mission critical that accelerate fatigue damage in the aircraft structure and may therefore recommend reducing such maneuvers.

The Canadian Forces also gather additional information on structural condition from the in-service inspection data and the results of any supplemental tests (e.g., tests on a failed component). The ASIP manager maintains inspection and maintenance data.

The CF-18 program office has transitioned to all-electronic data transfer and collection for both structural-use and maintenance data. Its maintenance database ties together all the data on maintenance activities, configuration changes, and structures (such as engineering drawings) for each aircraft using a software program called the Structural Information System. The CF-18 office uses this database to document and track all the maintenance and modification activities.

Because of the uncertainties associated with aircraft use and the engineering models, periodic assessments may be necessary to verify the structural condition of an aircraft during its service life. If that actual use of the aircraft has changed significantly from what was assumed when the aircraft was designed, it may be necessary to reverify its structural condition to substantiate its airworthiness. The WSM and the ASIP manager choose the method of compliance by proposing tests or analysis procedures, which the TAA must approve. For instance, both the CF-18 and CP-140 are undergoing SLAPs.[6]

---

[6] The Canadian Forces and the U.S. Navy share information on critical in-service experience, such as the discovery of unanticipated cracks on aircraft types they have in common.

These assessment programs involve full-scale fatigue tests of aircraft with a high number of flying hours, teardown inspections, and updating fatigue analyses. These allow verification of the structural conditions of these fleets and to identify areas for modifications that will extend their service lives.

Updates of ASIP Master Plan

The ASIP manager uses the current and predicted structural conditions (FLI, maintenance records, risk, etc.) and the requirements in the basis of certification to update the ASIP master plan as necessary. This plan outlines all the required structural-life management tasks for both the near and long terms, including updates to inspection, maintenance, and modification plans. If fleet-management plans (e.g., modifications) will significantly affect the aircraft operation, the operators are consulted during the decisionmaking process.

The frequency of ASIP master plan updates varies by weapon system. For example, the CF-18 ASIP Master Plan is updated continuously and is available online. The CP-140 ASIP Master Plan is updated annually.

The WSM authorizes and allocates funds for ASIP and fleet-management tasks (such as modifications); thus, the WSM must approve the ASIP master plan. The ASIP manager typically proposes the budget for the structural plan and assesses the operational and cost effects of the plan to assist the WSM's decisionmaking. For instance, the CF-18 ASIP manager uses a fleet-management software program called Aircraft Life Extension (ALEX) Planning Software to estimate aircraft availability, the costs of maintenance and modification, and the structural risk for various fleet-management scenarios.[7] These estimates of different scenarios help the ASIP manager propose an optimal plan to support the decisionmaking process.

During the SLAPs for the P-3C and F-18, these services shared a considerable amount of information.

[7] ALEX integrates the structural maintenance plan with the engineering data, maintenance and inspection data, modification data, operational constraints (e.g., flying rates, fleet size), and production constraints to estimate the cost, availability, and safety risks. The program was developed by L-3 Communications, Mirabel, Quebec.

The SLAP programs for both the CF-18 and CP-140 have been multinational efforts, which has mitigated some of the costs associated with these programs for the Canadian Forces.

Risk Management

The WSMs and the ASIP managers incorporate the risk-management process in their fleet-management framework. The risk-management process provides a framework for helping fleet managers decide how to balance flight-safety risks against operational effects (e.g., aircraft availability and mission capability) and costs.

The Technical Airworthiness Manual outlines a general technical airworthiness risk-management process similar to the U.S. Air Force's system risk-management framework, as shown in Figure 4.2.

The process begins with identification of hazards, such as cracks in wing root, followed by an assessment of the risk of mishaps due to those hazards. Risk-assessment methodologies vary by weapon system. Each weapon system program office is responsible, with TAA

Figure 4.2
Elements of Risk Management in Canadian Forces' Technical Airworthiness Program

approval, for establishing a suitable method of assessing risk. The CF-18 program office, for example, assesses structural risk in terms of the probability of catastrophic structural failure as a function of an aircraft's FLEI.

The WSM uses this assessment to identify and implement appropriate risk-control or risk-mitigation measures, such as increasing the inspection frequency. When implementing such measures, the WSM either accepts the residual risk or reports the risk items to appropriate authorities for accepting the risk. Finally, in the risk-tracking process, the WSM documents the risk items and verifies the effectiveness of any mitigation initiatives undertaken. Risk tracking is critical to risk management because it ensures that mitigation measures have been implemented or that the risk has been accepted. Risk tracking further provides traceability of risk-control outcomes. The risk-management process continues throughout an aircraft's service life, ensuring that risks are constantly monitored and managed.

Risk assessment enables the WSM to explore different resource allocation decisions by understanding their effects on structural integrity. For instance, the WSM can postpone a modification by choosing to use inspection methods to ensure aircraft safety for a limited period.

Regulatory Processes

The regulatory processes are essential to the Canadian Forces' aircraft structural-life management. The TAA ensures that structural-life management activities are conducted in conformance with the specified standards to achieve acceptable airworthiness via formal approval processes and formal program monitoring processes. These processes complement the ASIP process:

- airworthiness certification
- design-change certification
- risk acceptance
- periodic reviews and audits.

Airworthiness Certification

Each aircraft type is required to have an airworthiness certification before it is authorized to fly. For an aircraft type to be certified, the TAA must (1) verify that the basis of certification for the specific aircraft type is adequate and appropriate, (2) determine that the aircraft type has met the requirements in its basis of certification, and (3) grant airworthiness approval to the type design if compliance findings warrant.[8] The WSM can demonstrate that the aircraft type has met the requirements in the basis of certification through tests, analysis, or any other customized means that the TAA deems acceptable. The TAA then evaluates the compliance findings and grants flight authorization based on the airworthiness certification. Thus, the TAA can use flight authorization as a means of controlling risk and enforcing compliance.

Design-Change Certification

Before any plan that may affect structural integrity can be implemented, it must receive TAA approval via the design-change certification process. Any change in the aircraft's maintenance program, configuration, or mission constitutes a design change. The certification process ensures that no such alterations can compromise its airworthiness, including its structural integrity.

The process begins with the WSM's assessment of how the proposed design change would affect structural integrity. Depending on the results of this assessment, the design-change certification process may require steps similar to those of the airworthiness certification process described earlier. Because the initial basis of certification is only applicable to the original configuration and use, a design change requires preparation of a new basis of certification. The WSM develops this new basis, which the TAA must approve before airworthiness can be certified. The design-certification process requires the WSMs to examine airworthiness implications before implementing significant design changes and to formally track such changes.

---

[8] The Technical Airworthiness Manual describes steps of the airworthiness certification process in detail.

Risk Acceptance

The TAA establishes the acceptable level of technical airworthiness risk, tailoring it to each specific weapon system. For example, a cumulative probability of structural failure of 1 in 1,000 is deemed acceptable for the CF-18, while the acceptable risk level for the CP-140 is lower, 1 in 10,000. The TAA established a lower acceptable risk level because the CP-140's aircrew is larger.

The TAA further delegates the authority to accept a given level of risk to various personnel, such as the WSM. Increasing levels of risk, however, require increasing levels of authority. The highest risk level requires TAA approval, in consultation with the other authorities in the Airworthiness Program: the Operational Airworthiness Authority (OAA) and the Airworthiness Investigative Authority (AIA).[9]

The TAA consults with the OAA in accepting the highest technical airworthiness risks because operational effectiveness may take precedence over flight safety in certain instances, such as in wartime. If risk reduction requires immediate action, such as imposing flight restriction, TAA and the Chief of Air Staff evaluate the decision in terms of the effects of such an action on both the technical airworthiness and operational requirements.

Periodic Audits and Reviews

The TAA staff is very involved in providing oversight of the structural-life management process, informally monitoring ASIP compliance. The TAA staff may be involved in selecting and developing compliance methods for key ASIP processes (e.g., airworthiness certification, design certification, risk acceptance) and in helping organizations achieve compliance.

The Canadian Forces also incorporate formal program-monitoring processes, including the annual meetings of the Airworthiness Review Board (ARB) and periodic audits. The ARB consists of representatives from TAA, OAA, and AIA. During these meetings,

---

[9] The Minister of National Defence delegated the Commander of 1 Canadian Air Division as the OAA and the Director of Flight Safety as the AIA.

the board reviews the airworthiness status of all fleets and other airworthiness issues. The WSMs report risk items to the board. During its regular monitoring activities, TAA also reviews risk items and risk-control actions taken.

TAA recently audited all the fleet ASIP programs to evaluate their compliance levels. As a result of these audits, TAA required WSMs and ASIP managers to submit corrective action plans to bring themselves into compliance with current ASIP and technical airworthiness requirements. TAA also plans for ARB to review all fleet ASIPs annually to monitor compliance.

Summary

The characteristics of the Canadian Forces' approaches to aircraft structural-life management can be summarized as follows:

- Regulations
  - Policy is broad and is based on the concept of airworthiness.
  - Specific requirements are set for specific aircraft types.
  - An independent regulatory body, TAA, serves as the central authority.
  - Formal review and approval processes ensure compliance.
- Communications
  - Multiple types of information are used to convey structural condition.
  - The regulatory approach ensures communication, especially the sharing of critical information.
  - Critical information is formally communicated in the ASIP process.
  - Established working relationships and colocation facilitate informal communications.
- Resource management
  - Each fleet has a single funding authority (WSM).
  - Regulations and risk assessment guide prioritization of resource allocation.

- The ASIP master plan facilitates formal planning for resource management.
- TAA, the central regulatory authority, provides independent technical assessments for resource decisions.

Regulations

The Canadian Forces' governing policy regarding structural integrity is generally based on the concept of airworthiness, using the basis of certification as a means of assessing compliance. Because the basis of certification lists specific requirements for the specific aircraft type, the policy governing structural-life management for each aircraft type can be explicit.

The Canadian Forces have a central, independent regulatory body that regulates and oversees all aspects of aircraft structural-life management (both technical and programmatic). The strategy is to provide a balanced level of regulation and flexibility by assessing compliance case by case for many aspects of structural-life management.

TAA provides oversight via formal approval and monitoring processes to ensure compliance. The approval processes focus on the tasks that are the linchpins of aircraft structural-life management, such as airworthiness certification and design-change certification, to balance the level of regulation in this area.

Communications

For the Canadian Forces, an aircraft's structural condition and airworthiness are tied together. As a result, the Canadian Forces use multiple types of information to convey structural conditions (e.g., FLI, risk) because determining airworthiness, as far as structural integrity is concerned, requires meeting multiple requirements.

The regulatory processes require formal communication of a great deal of information. Especially, the processes require documentation of critical information for traceability and planning purposes, as well as for determining compliance.

In addition to the formal communications, informal communications between the WSMs, the ASIP managers, and the TAA staff occur during various decisionmaking processes because of their close working relationships. Colocation of these personnel further facilitates informal communications. Although ASIP implementation is organizationally decentralized, colocation also leads to informal information sharing among ASIP managers, providing visibility across fleets and cross-fertilization of ideas.

Resource Management

The WSM is the sole funding authority for structural-life management. Such regulatory processes as reviews, risk monitoring, and certification provide independent assessments of resource allocation decisions. The ASIP master plan provides formal resource-management planning.

Regulations and risk assessments guide prioritization of resource allocation. Regulations focus on critical ASIP tasks, such as monitoring structural conditions. The WSMs have some flexibility in managing their resources and may choose any compliance method, as long as TAA approves. Risk assessment helps the WSM compare alternatives to evaluate how different resource allocation decisions affect structural integrity.

CHAPTER FIVE

Aircraft Structural-Life Management in the U.S. Air Force

The U.S. Air Force operates about 6,000 fixed-wing aircraft of about 40 different aircraft types in a land-based environment (Air Force Association, 2004). The Air Force uses the damage-tolerance approach to manage aircraft structural life and to establish the maintenance plans for its aircraft.

This chapter describes the U.S. Air Force's ASIP. We describe the program's structure in terms of its governing policy and responsible organizations, followed by the principal elements of its processes. The chapter concludes with a summary of the program's approaches to regulation, communication, and resource management.

Program Structure

Table 5.1 lists the documentation governing the U.S. Air Force ASIP. The following paragraphs describe this documentation in more detail. Appendix A gives a history of the program and of its governing documentation.

Air Force Policy Directive (AFPD) 63-10, Aircraft Structural Integrity, requires establishment of an ASIP "for each aircraft weapon system it is acquiring or using," tailored to that system. As a result, ASIP implementation and execution are decentralized, taking place at the system program office (SPO) for an aircraft type or a mission, design, or series (MDS).

Table 5.1
Documents on ASIP Policy, Procedures, and Standards

Document	Type	Purpose
AFPD 63-10	Policy directive	Establishes ASIP policies and assigns broad responsibilities for implementing both the overall ASIP and individual programs
AFI 63-1001	Policy instruction	Establishes procedures and specific responsibilities for implementing both the overall ASIP and individual programs
MIL-STD-1530B	Military standard	Describes the overall ASIP program and provides technical direction for managing and carrying out individual ASIPs

The directive also assigns broad ASIP responsibilities to various organizations. The corresponding AFI, 63-1001, defines procedures for implementing and sustaining the program, as well as specific organizational responsibilities. ASIP is described in a military standard, MIL-STD-1530B, which provides technical direction for managing and executing the program. The military standard includes requirements for design, analysis, and test procedures to establish structural integrity during the design and development phase, as well as requirements for data collection, inspection, maintenance plans, and modifications for sustaining structural integrity during the sustainment phase. As mentioned in Chapter One, before February 2004, the ASIP requirements now in MIL-STD-1530B had been treated as guidelines and were listed in a military handbook (MIL-HDBK-1530B).[1]

Governing Policy
The governing policy behind the Air Force's approach to structural-life management is to prevent structural failures through effective

---

[1] As mentioned in Chapter One, the Air Force released MIL-STD-1530C in November 2005, as this report was being prepared for publication. These changes do not significantly affect our findings. Some of the key changes in MIL-STD-1530C are addition of (1) structural certification (and recertification when necessary) to ASIP as part of airworthiness certification and (2) risk analysis in ASIP tasks. Many of the changes may not yet be in practice because additional work will be required to implement the new requirements.

maintenance. This requires a maintenance plan that includes inspection for fatigue damage and timely repairs and modifications, resulting in cost-effective life-cycle management (AFPD 63-10).

For each aircraft type, the U.S. Air Force develops a Force Structural Maintenance Plan (FSMP), which sets out a schedule for performing the maintenance actions (inspection, repair, and modification) necessary for maintaining structural integrity throughout the fleet's service life. The FSMP provides a basis for inspection and maintenance requirements for individual aircraft, as well as for estimating the maintenance costs the fleet will incur during its service life.

The inspection program monitors fatigue damage (cracks) at critical locations on the aircraft to ensure that the accumulated fatigue damage does not reach the failure threshold (critical crack size) during the aircraft's service life. The U.S. Air Force tracks aircraft use to update the FSMP and inspection plans (when and where to inspect for fatigue cracks) to ensure timely detection and repair of fatigue damage in critical locations. The inspection results can drive further inspection, repair, or replacement of damaged components.

The U.S. Air Force replaces a fleet when continuing to maintain it becomes uneconomical or when its operational effectiveness has degraded. For example, rapid growth in the number of cracks in multiple fatigue-critical areas may require multiple major modifications that may in turn have significant implications for operational effectiveness and cost.

The directive also addresses ASIP compliance measures (AFPD 63-10, Attach. 1, para. A1.1, p. 3):

> Compliance with this policy directive will be assessed based on the number of Class A and B accidents due to structural problems. This is a valid indication of the health of any system. It can be derived from data maintained by the Air Force Safety Center. Other factors that should be considered include the age of the fleet, structural repair costs and frequency, operational readiness, etc.

A shortcoming of this policy is the fact that a metric based on the mishap rate—on actual structural failures—is a lagging indicator of ASIP compliance. This metric is thus not useful for proactive ASIP management.

The ASIP for each MDS may use additional measures of compliance. For instance, the concept of "economic life" has been developed as a measure of cost-effectiveness of structural-life management. Although the SPO may use the FSMP to estimate the expected economic life for its fleet, the Air Force has not adopted a standard method of computing economic life. As a result, this is not used as a standard measure across the Air Force.

Principal Authorities

Figure 5.1 shows the Air Force organizations responsible for structural-life management. A dotted line indicates a supporting relationship; a solid line indicates command relationship.

The Assistant Secretary of the Air Force for Acquisition (SAF/AQ) is responsible for ensuring that an ASIP is established for

Figure 5.1
ASIP Responsible Organizations in the U.S. Air Force

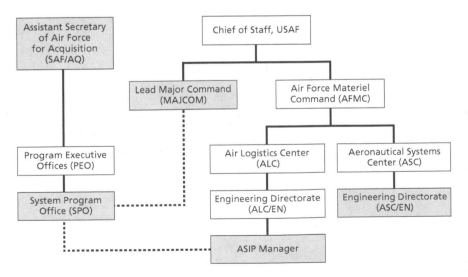

each MDS. Each SPO is responsible for ensuring that its ASIP continues throughout its MDS's operational life. The head of the SPO, the system program director (SPD),[2] appoints an ASIP manager to carry out the program for his or her weapon system, in keeping with the AFI. The ASIP manager establishes the program, tailoring it to the aircraft type in accordance with MIL-STD-1530B. The SPD must approve the resulting ASIP.[3]

Within the U.S. Air Force, four major commands (MAJCOMs) operate aircraft to meet force requirements: the Air Mobility Command (AMC), the Air Combat Command (ACC), the Air Force Special Operations Command (AFSOC), and the Air Education and Training Command (AETC). When one or more of these MAJCOMs possesses the same type of weapon system (AFPD 10-9), one of them will be designated as the lead command. A single command may have multiple fleets, which will have commonalities in capabilities and use. For instance, ACC owns fighters and bombers (weapon systems), while AMC owns transports and tankers.

The lead command for any fleet is the advocate for that fleet. It establishes and prioritizes sustainment requirements to meet operational requirements. The SPDs support the lead command in developing the fleet-management plans and in executing them. The lead command has the funding and decision authority over scheduling fleet sustainment activities, including ASIP tasks. As a result, the lead command has a strong influence on the effectiveness of its fleets' ASIPs.

The policies for ASIP (policy directives, instructions, and standards) have been primarily established by SAF/AQ and by the Air Force Materiel Command (AFMC). AFMC has, in turn, designated ASC/EN as the office of primary responsibility (OPR) for ASIP.[4] In

---

[2] U.S. Air Force publications use both *SPD* and *single manager*, depending on when the document was drafted. Since the terms are synonymous, we use only SPD.

[3] MIL-STD-1530C requires every ASIP program to address all tasks and elements in the standard and that, if any task or element in the standard is to be tailored, the effects of that tailoring must be documented in the ASIP master plan and must be approved.

[4] In accordance with AFI 63-1001, AFMC Supplement 1, April 2003.

this role, ASC/EN advises on technical policy and procedures and provides oversight for the program.

Because of the sheer size of the Air Force's organization, those holding ASIP responsibilities are geographically dispersed, as shown in Table 5.2. The lead commands, ASC/EN, and SAF/AQ are spread across the country. The SPDs and ASIP managers (for aircraft that are no longer being procured) operate at one of three air logistics centers (ALCs), depending on the particular MDS.[5] This arrangement is partly due to the fact that the ALCs focus on aircraft sustainment, while ASC focuses on aircraft acquisition. Colocating SPDs at an ALC facilitates communications among the engineering staff, the logisticians, and the maintainers in responding to new maintenance issues as they emerge. Each ASIP manager is part of the ALC's chain of command, under the engineering branch (ALC/EN). The corresponding SPD is part of a separate chain of command, reporting to the PEO.

ASIP Process

The following are the principal elements of the U.S. Air Force's aircraft structural-life management program:

- developing ASIP master plans
- developing FSMPs
- tracking structural use
- updating FSMPs (continuous)
- implementing FSMPs and ASIP master plans.

The following subsections address each element.

---

[5] SPDs for the aircraft that are in the acquisition phase or still in the procurement process (e.g., C-17) are stationed at either Washington, D.C., or Wright-Patterson Air Force Base (AFB), Ohio.

Table 5.2
Geographic Dispersion of ASIP-Responsible Organizations

Organization	Role	Location(s)
SPDs	Fleet-management support	Warner Robins ALC, Georgia Oklahoma City ALC, Oklahoma Ogden ALC, Utah
ASIP managers[a]	ASIP program management	Warner Robins ALC, Georgia Oklahoma City ALC, Oklahoma Ogden ALC, Utah
ACC	Lead command	Langley AFB, Virginia
AETC	Lead command	Randolph AFB, Texas
AFSOC	Lead command	Hurlburt Field, Florida
AMC	Lead command	Scott AFB, Illinois
ASC/EN	Engineering	Wright-Patterson AFB, Ohio
SAF/AQ	Acquisition system supervision	Washington, D.C.

[a]Note that the ASIP managers are assigned to same centers as their respective SPDs.

Developing ASIP Master Plans

Very early in the life cycle of an aircraft, the ASIP manager implements the requirements in AFI 63-1001 to create an ASIP program for the designated MDS, documenting it in an ASIP master plan. The master plan contains the scheduling and spells out the ASIP tasks that must be carried out throughout the aircraft's life cycle to achieve and maintain structural integrity, following the directions provided in MIL-STD-1530B.

MIL-STD-1530B describes the main ASIP tasks (see Figure 5.2). ASIP tasks I through III focus on activities required during the acquisition phase, including establishing structural integrity design requirements and developing and executing analysis and test programs to determine compliance with the requirements. Tasks IV and V focus on activities required to sustain structural integrity throughout an aircraft's life cycle, including developing in-service data-collection programs and maintenance plans.

The ASIP manager is responsible for coordinating the specific ASIP with other responsible organizations, such as the lead command, and for keeping them informed of the initial plans for and any

Figure 5.2
Main ASIP Tasks Outlined in MIL-STD-1530B

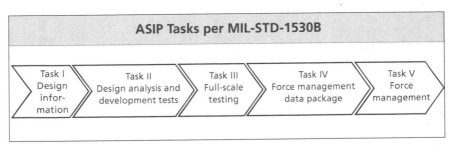

NOTE: Task IV of MIL-STD-1530C includes aircraft structural certification. The results of tasks I through III provide a basis for aircraft structural certification.

changes to the program. However, there are no formal processes to ensure such communication.

Before February 2004, the frequency of ASIP master plan updates was left to each SPO's judgment. MIL-STD-1530B now requires updates at least annually, more often if necessary (e.g., when changes in structural use affect structural integrity).

Developing Force Structural Maintenance Plans
An FSMP is a schedule for the maintenance activities necessary for monitoring and repairing fatigue damage to aircraft throughout a fleet's service life. It estimates the costs of these actions, whenever possible. Thus, a fleet's FSMP is a key management element, useful for maintenance, budgetary, and retirement planning (based on costs). FSMPs provide the following information:

- critical locations and crack sizes that require maintenance action
- anticipated inspection, repair, and modification actions
- supporting data required for developing the maintenance plan (e.g., initial test results during design and development).

The FSMP development process is described in MIL-STD-1530B and is illustrated in Figure 5.3. ASIP Tasks I through III provide the necessary data for predicting the crack growth for critical locations in the aircraft. The initial tests (full-scale fatigue tests,

Figure 5.3
FSMP Development Process

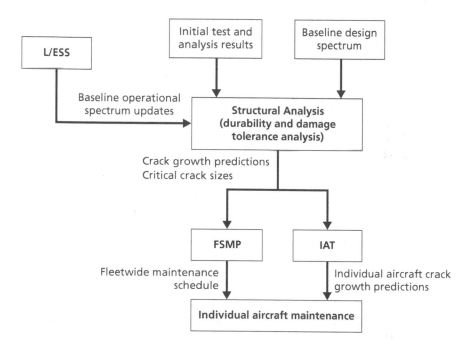

component tests, static tests, etc.) and the structural analyses (durability and damage-tolerance analyses) identify fatigue-critical locations in the aircraft, crack growth rates, and the critical crack sizes at which failure would occur. These data assume a specific load spectrum for the operational aircraft. The maintenance (inspection and repair) schedule is then based on the predicted crack growth curves and critical crack sizes.[6]

The crack growth curves and the corresponding maintenance times are based on the baseline, or the average, load spectrum for the fleet. The maintenance plans for specific aircraft take into account differences in how individual aircraft are used. The individual maintenance plan includes a schedule for each critical location of the aircraft.

---

[6] For more on the damage-tolerance method, see Chapter Two.

Almost always, the actual load spectrum aircraft undergo is different from what they were designed for. As a result, the initial FSMP needs to be updated using the baseline spectrum that represents the actual use; this is referred to as the operational spectrum. The Air Force uses a tracking program called the Loads and Environmental Spectra Survey (L/ESS) to collect flight data from which to compute the baseline operational spectrum for the fleet (Figure 5.3). An updated structural analysis (durability and damage-tolerance analyses) using the operational spectrum provides new crack growth predictions, which are then used to update the FSMP and the individual maintenance plans. Similarly, the maintenance plan for an individual aircraft needs to be updated using the actual use data for that aircraft. These data are provided by the Individual Aircraft Tracking (IAT) program.

Tracking Structural Use

Two programs help the Air Force track aircraft structural use: L/ESS and IAT. As just discussed, L/ESS helps determine the fleetwide baseline operational spectrum. The program monitors the actual use of the aircraft during its first few years of operation. A sample of the fleet is instrumented to collect the flight data. The data-collection requirements for L/ESS vary by MDS. The ASIP manager must select a statistically meaningful sample size, which will depend on the fleet's size, for the L/ESS program. The sample size is typically about 15 to 20 percent of the fleet, and the ASIP standard recommends collecting data for five years.

The IAT program tracks individual aircraft by tail number to monitor their variation from the fleetwide baseline throughout their service lives. Any significantly different use would be captured, and the individual aircraft maintenance plans would be updated accordingly.

IAT data-collection requirements vary for each weapon system. For example, transports and bombers tend to vary less from their fleetwide baseline than fighters do from theirs. The SPO may continue to collect L/ESS data throughout the service life to detect any changes in the baseline spectrum. Alternatively, if the IAT data are sufficient to compute the baseline spectrum, L/ESS may not be necessary.

The ASIP manager is responsible for planning and managing both the L/ESS and IAT programs, which involve tailoring a tracking analysis method for the MDS and defining the corresponding data-collection requirements, such as types of data and sample sizes. The ASIP manager and the SPD determine the frequency of data analysis to ensure timely collection and analysis to facilitate updates to the maintenance and inspection plans and the FSMP.

The IAT and the L/ESS data are collected at a centralized database maintained at the Oklahoma City ALC. The methods and data systems used to report the data to the centralized reporting facility vary by MDS because data-collection plans have been customized for each MDS.[7]

Each SPO analyzes the data for its own fleet to (1) predict crack growth for each critical location on each aircraft and (2) adjust the maintenance plans as necessary. The structural-use tracking programs provide continual assessment of the aircraft's structural condition and a basis for adjusting its maintenance plans as necessary to sustain structural integrity. In most cases, the ASIP manager uses contractors to support such data and engineering analyses.

Updating FSMPs

To ensure that an effective maintenance program is in place to prevent structural failure, the FSMP needs continual assessment and validation. As Figure 5.3 illustrates, the validity of the FSMP depends on the accuracy of the load spectrum and the structural analysis (crack growth prediction). Each SPO is responsible for collecting adequate L/ESS and IAT data to update the structural use tracking as necessary.

MIL-STD-1530B requires an FSMP update if the use or the configuration of the aircraft changes significantly during its service

---

[7] While the data for most of the fleets are stored at Oklahoma City ALC, a few aircraft report their ASIP data to other locations. Data for C-130s and E-8s go to Warner Robins ALC, and data for B-2s, F-117s, and T-38s go to contractors.

life. Increasing the payload, for example, would be a change in the aircraft use. A configuration or mission change would alter the stresses on the structure, thus making it necessary to update the baseline spectrum and structural analysis. The lead command, as the operator, is responsible for communicating any changes in aircraft use to the SPD so that their effects on structural integrity can be assessed properly.

An FSMP update may be required if crack growth revealed during in-service inspections differs significantly from the predictions. An in-service inspection could reveal unanticipated cracks in a new critical area or accelerated crack growth, potentially indicating inaccuracies in the damage-tolerance analysis. Reassessments of the model and analysis may also be required to ensure that the updated FSMP is accurate. Each SPO is responsible for keeping the FSMP up to date and for determining the adequate level of tasks for updating and validating it. The level of effort can vary from simple data analysis to a new full-scale fatigue test and teardown (similar to the U.S. Navy's SLAP).[8]

The 1997 NRC report mentioned in Chapter One recognized the importance of an up-to-date FSMP and recommended reassessments every five years, especially for the older fleets (Tiffany et al., 1997). However, updates to FSMP have primarily been driven by the available level of ASIP funding.

Implementing FSMPs and ASIP Master Plans

The final step in the ASIP process is execution of the maintenance actions set out in the FSMP and the tasks specified in the ASIP master plan, as scheduled. Timely execution of these tasks enables proactive sustainment, a cost-effective way to prevent structural failure. The SPD is responsible for an adequate implementation of the FSMP and the ASIP master plan, with funding from the lead command.

---

[8] MIL-STD-1530C requires recertification of the aircraft structures if configuration, aircraft structural use, damage, or service-life expectancy is significantly different from what was assumed in aircraft certification.

Before 1997, SPD had held budgeting and funding authority for ASIP. The revision of AFPD 63-10 moved this authority to the lead command, thereby giving the command authority over ASIP tasks and the scheduling of fleet sustainment decisions (e.g., maintenance, modifications). The lead command was given this function because it is the operational user of these aircraft and provides forces to the combatant commanders during crises. The lead command can weigh combat commitments against sustainment commitments in making the best decision.

The annual budgetary—Program Objective Memorandum (POM)—process, is the formal outlet through which the SPD communicates structural sustainment needs (e.g., ASIP tasks, maintenance actions) to the lead command.[9] The ASIP manager evaluates fleet needs in terms of maintaining structural integrity and builds the requirements based on the FSMP and the ASIP master plan. Examples of such requirements are

- procuring and installing improved data-collection systems (i.e., newer recorders or automated data downloads)
- adding new sensors (e.g., strain gauges to gather additional data as fleets age)
- programming for structural modifications or component replacements to ensure that the MDS can meet its designed service life
- programming for additional inspections or maintenance work during programmed depot maintenance
- programming for a new full-scale fatigue test.

The SPD evaluates such program elements for ASIP, as well as other program elements required for sustaining the fleet, and makes the necessary trade-offs to construct a budget program for fleet sustainment. The SPD forwards the resulting POM input to the lead

---

[9] The POM process is a part of the programming phase of the DoD budget process, in which each DoD component develops a POM and submits it to the Office of the Secretary of Defense for review. The POM entails a six-year funding plan to accomplish overall program goals and milestones.

command, which reviews the requirements further in the context of meeting the operational requirements. The lead command then considers the trade-offs between operational modifications (e.g., an improved radar) and sustainment modifications (e.g., repairing corroded fuel tanks). The POM submissions for MDSs within a single lead command are also balanced against each other.

ASIP budgetary programming can be challenging for the lead command for a couple of reasons. First, the lead command does not have the expertise in ASIP and structural needs. Second, it is difficult to compare the relative needs of the different fleets within a command because of the varying methods of measuring structural condition the SPOs use for their own MDSs. For example, crack sizes and the crack density (number of crack locations) on an aircraft represent accumulated fatigue damage and are related to the structural condition. However, there is no standard method of converting such data to a metric that would convey these conditions to others, and there is no standard threshold that would convey what constitutes an acceptable crack distribution. The assessment of structural conditions is thus left to the judgment of the ASIP manager and the SPD.

Some lead commands (AMC and AFSOC) have their own OPRs for ASIP to facilitate communications with the SPDs regarding fleet-management decisions and ASIP tasks. Both AFSOC and AMC have the advantage of an engineer assigned as the ASIP OPR. These ASIP OPRs intimately understand and can advise their lead commands on the technical issues. The OPR can also go back to the SPDs to explain the operational effects that certain fleet-management plans and ASIP tasks would have on the command and try to develop more-feasible ways to meet the fleet sustainment requirements. Involving an engineer in this process allows the lead command to make decisions with more in-depth knowledge across the MDSs and to build more effective trades with the SPDs.

Ultimately, the lead command reviews the POM inputs from each SPD within its command and then balances the operational needs against the structural integrity needs across multiple fleets to allocate the resources.

Risk Management

The Air Force's program risk-management framework is based on MIL-STD-882D, Standard Practice for System Safety. The framework includes hazard identification, risk assessment, risk mitigation, and risk acceptance. As with the Canadian Forces and the U.S. Navy, the distributed risk-acceptance structure requires increasing levels of risk to be addressed by increasingly higher authorities.

Each SPO uses a risk-management approach tailored to its system. Currently, there is no standardized risk-assessment method or standard level of risk acceptance for aircraft structures. The Air Force is in the process of institutionalizing a risk-assessment framework in ASIP under the guidance of the senior advisor, ASC/EN.[10] This is a joint effort between industry and the Air Force to establish a framework for structural risk assessment that can be integrated in the context of the MIL-STD-882D. The main challenge in applying this framework for structural risk assessment is quantifying the probability of structural failure. Recent ASIP conferences and a risk-assessment workshop have addressed ways to implement probabilistic methods in structural analyses to quantify structural failure probabilities.[11] While the Canadian Forces use a deterministic risk-assessment method, their experience may offer some insights for the U.S. Air Force effort.

Regulatory Processes

There are no formal reviews of ASIP or regulatory processes that monitor compliance with ASIP policy and standards. According to

---

[10] MIL-STD-1530C has made a progress in this effort and now requires a risk analysis using an estimate of initial flaw size distribution, although a specific risk analysis method is not discussed. The standard further defines acceptable and unacceptable risk levels for military operations as probabilities of catastrophic failure at or below 10^{-7} per flight and exceeding 10^{-5} per flight, respectively.

[11] The 2004 ASIP Conference in Memphis, Tennessee, and the June 2004 Risk-Based Force Management Workshop in Dayton, Ohio.

the AFPD and the AFI, regulatory responsibilities have been assigned to SAF/AQ, AFMC, and the SPDs.

AFMC is responsible for ensuring that up-to-date ASIPs are implemented for all aircraft systems. ASC/EN, as AFMC's OPR, provides oversight for this, establishing procedures for periodic reviews of each fleet's program. These reviews are essential to ensuring that the decentralized programs adhere to the prescribed directives. The reviews should verify that the data collection and analysis are effective and efficient and that ASIP inspections are valid and on schedule.

There is a discrepancy between the AFPD and the AFI on ASIP approval authorities. According to the AFPD, SAF/AQ approves ASIPs. In the AFI, each SPD approves the ASIP for its own MDS. This inconsistency may be due to the policy directive's focus on ASIP during the acquisition phase and the instruction's focus on the sustainment phase. The Air Force is currently revising both documents to clarify ASIP approval authority.

Although these regulatory responsibilities have been assigned, ASIP has not been strictly enforced, partly because relevant standards were guidelines before February 2004 and thus not enforceable. SAF/AQ, ASC/EN, and the SPDs had no authority over the lead 'commands' decisions on ASIPs. As a result, compliance with ASIP policies has been primarily self-managed, by individual SPOs. With the return of the standard as an enforceable requirement, control of ASIP is expected to be reestablished.[12]

Summary

The characteristics of the U.S. Air Force's approaches to aircraft structural-life management can be summarized as follows:

___

[12] MIL-STD-1530C defines *aircraft structural certification* and defines the ASIP tasks that establish structural certification.

- Regulations
  - The policy is broad and flexible and is based on broad objectives.
  - The program structure is flexible and decentralized, with minimal regulation and oversight.
- Communications
  - The measures that communicate structural condition are SPO-dependent.
  - There are no standard methods for communicating the health of various fleets.
  - Communications with decisionmakers about ASIP and structural issues are limited and informal.
  - Communication is, however, facilitated by the lead command's ASIP OPR.
- Resource management
  - The lead command acts as the single funding authority for structural-life management for its multiple fleets.
  - FSMPs and ASIP master plans guide budget inputs.
  - The commandwide view of structural needs is limited.
  - The regulations guiding resource prioritization are limited.

Regulations

The U.S. Air Force's governing policy on structural integrity is broad to allow an ASIP to be tailored for each MDS. The broad policy recommends multiple measures for assessing compliance, including the rates of mishaps due to structural failure. This measure, however, is problematic because it is a lagging indicator.

The U.S. Air Force has a flexible, decentralized regulatory structure with minimal ASIP regulation and oversight. As a result, the lead command has the strongest influence over ASIP implementation because it holds the funding authority.

Communications

Because of the decentralized ASIP implementation, the U.S. Air Force does not have a standard metric for communicating the structural condition of an aircraft. Each SPO applies its own technical judgment to assess the structural conditions of aircraft, using the FSMP, structural use tracking, and maintenance data on crack locations and crack growth to inform decisions.

The communication process is primarily informal, on an as-needed basis. It is each organization's responsibility to share the applicable information to achieve ASIP objectives. Communications between the SPD and the lead command about ASIP and structural conditions are limited because of the limited involvement of the lead command in the ASIP process, other than budget programming. There is no formal process ensuring regular communication between the SPD and the lead command. The geographic separation of responsible organizations in the Air Force further limits communications between the SPDs and the lead command, as well as among the different MDS ASIP managers within a command.

Resource Management

The lead command is the funding authority for managing its multiple fleets, including funding for their ASIPs. The lead command uses the proposed budget inputs from the SPDs, derived from the FSMP and ASIP master plan, to guide budget programming for the command.

The lead command lacks a commandwide view of the ASIPs and the structural conditions of aircraft because of ASIP decentralization. Some lead commands rely on their ASIP OPRs to better understand the implications of proposed tasks (e.g., risk of structural failure, implications for operational effectiveness, preventing costly repairs) to assist in the prioritization of resources.

There is no regulation to enforce certain ASIP tasks or to provide an independent technical assessment on the lead command's decision affecting structural integrity. Additionally, with no standard measure for ASIP compliance, there is no guidance on what would be an adequate level of resources for an effective ASIP.

CHAPTER SIX

Observations About Aircraft Structural-Life Management Approaches

In previous chapters, we described how the aircraft structural-life management programs of the U.S. Navy, the Canadian Forces, and the U.S. Air Force have approached program regulation, communication, and resource management. In this chapter, we discuss the implications of these approaches for enabling independent and balanced regulation, clear and timely communications, and adequate and stable resources.

Approaches to Regulation

We have characterized ASIP regulation approaches in terms of the governing policy, the oversight approach, and the distribution of regulatory responsibilities in each service. Table 6.1 compares these characteristics for the three services we examined.

Governing Policy

Making policies explicit ensures that they will be clearly understood, including what the acceptable compliance level is. The use of a quantifiable metric and a threshold simplifies compliance assessment. Additionally, the use of a common standard metric (e.g., FLE) for all aircraft types provides a forcewide view of compliance with the policy. Explicit policies, however, can limit decisions in structural-life management, such as determining when to retire an aircraft, and leave little room for tailoring to a specific weapon system.

Table 6.1
Comparison of Characteristics in Regulations Approach

Service	Characteristics
U.S. Navy	Structural-life management policies are explicit. Use of FLE as a metric provides visibility to ensure compliance. Authority for the technical aspects of structural-life management is centralized.
Canadian Forces	The policy is broad and is based on the concept of airworthiness. Specific aircraft types have specific requirements. The regulatory structure is independent but organizationally centralized. Formal review and approval processes exist to ensure compliance.
U.S. Air Force	Policies are broad and flexible and are based on broad objectives. The regulatory structure is flexible and decentralized, with minimal regulation and oversight.

Broader policies provide some flexibility in structural-life management to focus on additional objectives, such as cost-effectiveness, but at the expense of a case-by-case assessment of compliance.

Oversight Approach

Because the U.S. Navy's policies are explicit, the oversight of its structural-life management is minimal. Visibility of the compliance metric (FLE) enables organizations to monitor their own compliance with the policy, simplifying the management process. Lack of oversight, however, raises the risk of inadequate compliance with the policy and requirements.

Formal monitoring and approval processes provide the oversight for the Canadian Forces. These regulatory processes ensure that all the ASIPs are compliant but can add complexity and can require additional resources.

As with the U.S. Navy, ASIP oversight is minimal in the U.S. Air Force. However, unlike the U.S. Navy, the U.S. Air Force has a broad policy with no standard measure of compliance. This risks inadequate ASIP implementation because of the lack of clarity about the acceptable level of ASIP compliance.

Distribution of Regulatory Responsibilities

Both the U.S. Navy and the Canadian Forces assign regulatory authority for structural-life management to a single organization (NAVAIR and TAA, respectively). Centralizing regulatory responsibilities to a single organization leads to fuller visibility of all fleet's structural integrity compliance. It also provides clarity about who has authority for various aspects of the process.

The U.S. Air Force decentralizes regulatory responsibilities across multiple organizations (SAF/AQ, ASC/EN, and SPD). Coordination between these organizations can become challenging, leading to poorer visibility of compliance. Centralizing regulatory responsibilities, however, can pose a risk that the authority will gain too much control over structural-life management. Decentralizing the responsibilities provides checks and balance among regulatory authorities.

It is important for the regulatory organization to remain independent to provide such checks and balance. The U.S. Navy's explicit policies, along with its centralized program for structural-life management (ASLS), promote independent assessments of structural integrity from the PMAs. But a centralized structural-life management program limits tailoring.

The Canadian Forces have a separate regulatory organization (TAA) whose function is to regulate and oversee ASIPs, thus providing independent regulation.[1] This approach can complicate the management process and may require more resources. The Canadian Forces may be able to manage such an approach because of its relatively small force.

Some of the U.S. Air Force's ASIP regulatory responsibilities are assigned to organizations that are not directly involved in the management or execution of individual ASIPs during sustainment (ASC/EN and SAF/AQ). As a result, independent assessments of the programs are possible. However, the lack of expertise in the sustainment aspects of ASIP could make compliance assessments inadequate.

---

[1] Recall from Chapter Three that ASIP managers now support the WSMs solely and are not part of the TAA staff.

Approaches to Communication

We examined the types of information that are used to communicate the structural conditions and the information-sharing mechanisms, including the role of regulation. Table 6.2 summarizes these characteristics.

Types of Information Shared

The use of a quantifiable metric (e.g., FLE) and setting a threshold for that metric (e.g., FLE of 100 percent) make understanding the structural condition of aircraft simple and clear. Use of a standard metric for every aircraft and aircraft type provides a forcewide view on the structural conditions and life limits of that force's aircraft. As a result, it is easy for decisionmakers to understand the relative states of the different aircraft and aircraft types for fleet-management planning (e.g., modification priorities, scheduling, replacement plans).

Table 6.2
Comparison of Characteristics in Communications Approach

Service	Characteristics
U.S. Navy	A single, standard metric, FLE, conveys structural conditions.
	The results of rigorous fatigue-life tracking are disseminated frequently, through a formal fatigue-life report.
	Close working relationships and colocation promote and facilitate informal communication.
Canadian Forces	Multiple types of information are used to convey structural conditions.
	Regulations exist to ensure communication and sharing of critical information.
	Critical ASIP information is communicated formally.
	Colocation and close working relationships facilitate informal communication.
U.S. Air Force	The measures that communicate structural condition are SPO-dependent.
	No standard method exists for communicating the health of various fleets.
	Communications with decisionmakers on ASIP and structural issues are limited and informal.
	The lead command's ASIP OPR facilitates communication.

Developing a standard metric may be more complicated for aircraft that are managed under the damage-tolerance approach. In damage tolerance, structural condition is not measured just as a function of how the aircraft has been used, as it is in the calculations of the FLE. The maintenance program is also a key factor. As discussed in Chapter Two, the damage-tolerance philosophy tolerates cracks for a specified period, and they are repaired when detected. Depending on how each aircraft is inspected and repaired (or not repaired), two aircraft used identically may be in different conditions. Thus, any assessment of structural condition must take actual (as opposed to planned) maintenance activities into account. As a result, better understanding of the structural conditions of the force as a whole will require multiple types of information.

The lack of a standard metric that conveys structural condition is more of a problem for the U.S. Air Force than for the Canadian Forces because lead commands in the U.S. Air Force need to understand the relative states of multiple fleets and to gain a commandwide view for resource decisions.

A centralized program that monitors the structural conditions of all aircraft (e.g., SAFE program) would promote standardization in data collection, data analysis, technical methods, and dissemination of information for all aircraft types. Such a program would also facilitate the development and use of a standard metric for conveying structural conditions.

Such standardization does have limitations, in that it does not allow tailoring to a specific aircraft type. For example, fighters tend to have larger variations in use than do transports. As a result, frequent collection of data on flight loads is more critical for fighters than for transports. What constitutes an adequate frequency of data collection can vary significantly between different types of aircraft. Thus, a standardized approach could mean that data are collected either too often or not often enough.

A standardized approach can also result in a standard risk acceptance level for all aircraft types, as in the U.S. Navy. Again, depending on the aircraft type, such an approach could lead to a level that is too high for some aircraft and too low for others.

Information Sharing Mechanisms

Regulation of structural-life management should include formal communications (i.e., required communication as part of a formal process, meeting, or review) to ensure sharing of critical information with appropriate authorities. Formal communication typically involves documentation. Written communications are helpful for traceability and for future analyses, as well as for facilitating wide distribution of information across multiple organizations.

Formal processes promote regular information-sharing, which enhances timeliness. Formalizing much of the communications, however, can require additional resources and can introduce complexities that lead to inefficiency (such as by becoming too bureaucratic). Dependence on regulations to control communications can also restrict the types of information that are shared and can limit the information flow.

Informal communications rely on the responsible organizations to share the applicable information, leaving the types of information shared and the information flow flexible. This flexibility encourages open communication and free flow of information. At the same time, there is a risk that some communications may be missed or that some critical information may be delayed.

As with the Canadian Forces and the U.S. Navy, establishing close working relationships and colocating key personnel may facilitate informal communication. Such working arrangements promote regularity in information sharing, forcewide visibility, and some standardization in practices.

Approaches to Resource Management

Resource allocation, the final step in the ASIP process, authorizes the ASIP tasks. The resource-allocation step involves balancing safety and operational risks against costs to meet multiple needs for multiple aircraft and fleets.

Table 6.3 summarizes the resource management approaches that the U.S. Navy, Canadian Forces, and U.S. Air Force have taken. In this section, we discuss the implications of funding mechanisms and the roles of regulations and communications in resource management.

Dedicated Funding Line

Having dedicated funding for a particular task ensures that the task is carried out and at the proper time. It also provides stability for that task. However, dedicated funding could limit flexibility by preventing reallocation of resources to other activities that might turn out to have a higher priority from an overall fleet-management perspective. Thus, it is important to apply the dedicated funding to a task or a set of tasks whose stability is critical to ASIP viability.

Degree of Regulation

Explicit regulations ensure adequate and stable funding for specific tasks, but such restrictions may not be the most cost-effective. Regu-

Table 6.3
Comparison of Characteristics in Resource Management Approach

Services	Characteristics
U.S. Navy	The structural-life tracking program has dedicated funding.
	Use of FLE helps prioritize resources.
	The explicit policies drive key resource-allocation decisions.
	Independent technical assessments also play a part in resource allocation.
Canadian Forces	A single authority (the WSM) controls funding for structural-life management for the designated fleet.
	Regulations and risk assessment guide prioritization of resources in planning.
	The ASIP master plan provides formal planning for resource management.
	A regulatory body (TAA) exists to provide independent technical assessments for resource decisions.
U.S. Air Force	A single authority (the lead command) controls funding for structural-life management of the multiple fleets in the command.
	The FSMPs and ASIP master plans guide budget inputs.
	The commandwide view of structural needs is limited.

lations restrict flexibility in resource allocation, potentially compromising resources needed for more-urgent or higher-priority tasks.

A broad regulation structure provides broad guidance on prioritization of resources and leaves specific judgments about cost-effective resource allocation to the appropriate funding authorities. The risk is that such an approach could lead to lack of clarity about what constitutes an adequate level of resources for ASIP. Thus, it is important to choose tasks that are fundamental to ASIP viability, provide stability for these tasks, and avoid the risk of overregulation.

Involving an independent technical authority (e.g., NAVAIR and TAA in the U.S. Navy and the Canadian Forces, respectively) in resource management decisions can ensure that they do not compromise aircraft structural integrity. Such an authority provides checks and balances from a technical perspective and thus can enable adequate ASIP funding.

It is, however, possible to assign too much regulatory control to the technical authority. Allowing technical needs to overpower other critical needs in fleet management could interfere with the operational effectiveness of the fleet. The technical regulatory authority should be balanced to ensure adequate and appropriate ASIP implementation, while minimizing the risk of inadequate funding for other critical fleet-management needs.

Types of Information Used in Resource Management

We have already discussed the implications of using a quantifiable, standard metric to communicate structural conditions and to gain commandwide and forcewide visibility. Such information provides understanding of relative conditions across aircraft and fleets, facilitating prioritization and scheduling of resources. A metric that decisionmakers can use effectively for resource allocation would further emphasize the importance of such ASIP tasks as data collection and analysis, which are required to track such a metric. Using a leading indicator (e.g., FLE) would enable a decisionmaker to be more proactive and better informed when allocating resources. This is more effective than retroactively looking at mishap information to measure compliance.

Communicating the consequences of the resource decisions for ASIP objectives (safety, cost-effectiveness, and operational effectiveness) is also critical in guiding the prioritization of resources. This is one of the most difficult challenges in resource management. Certain methods, such as risk-assessment methods, can help quantify the consequences. However, it is important to establish a threshold so that the decisionmakers can clearly understand the acceptable levels of risk for achieving their ASIP objectives.

The U.S. Air Force's limited regulation of resource management and limited commandwide visibility may lead to inadequate ASIP funding over time. The return of the ASIP standard (MIL-STD-1530B) and the continuing development of a risk-assessment framework could mitigate this risk.

Formal Planning Processes

A formal resource planning process requires additional analysis and resources but enables proactive sustainment strategies and provides checks and balances for resource planning. A formal process (e.g., Canadian Forces' ASIP master plan update and approval process) should involve assessment of structural conditions, identification of necessary actions, a review of the plan, and documentation of final resource planning.

The documentation provides visibility of and traceability for resource allocations, enabling a better understanding of the future implications of decisions. Documenting and tracking the level of resources committed to sustainment of structures (including ASIP) would also improve future resource management.

Summary of Observations

Explicit ASIP policies provide clarity for determining compliance but limit flexibility in structural-life management. Broad ASIP policies, on the other hand, enable flexibility in ASIP implementation, allowing tailoring, but carry a risk of unclear understanding of what constitutes an acceptable level of ASIP compliance. Policies should be sufficiently explicit to provide general guidance about ASIP compliance but should also rely on independent, case-by-case assessments of ASIP compliance to enable tailoring.

ASIP regulations can provide checks and balances for structural-life management, enable clear and timely communication, and promote stable and adequate resources for the program. Regulations can also complicate processes and lead to management inefficiencies. The regulations should thus focus on ASIP elements that are critical to the program's viability and that ensure a balance between control and flexibility.

Organizational centralization would enable standardization of ASIP management and provide a forcewide view of ASIP compliance and the status of fleet. Decentralization would enable tailoring the program to fit specific weapon systems, achieving cost-effectiveness. Centralizing a set of selective ASIP tasks, where standardization is useful, could still allow other aspects of the program to be tailored for cost-effectiveness.

Approaches to regulation, communication, and resource management are highly interdependent and need to complement each other and the context of the program for the ASIP to be effective. Operational factors, such as force size, may present certain scalability challenges. The U.S. Air Force's large-scale force and wide range of aircraft types may pose some challenges for standardizing and/or centralizing certain aspects of ASIP across the force.

CHAPTER SEVEN

Observations About Options for the Future

Summary of Approaches

We drew the following observations from our survey of the three services' aircraft structural-life management programs:

- The U.S. Navy takes a rigid approach, managing aircraft structural life through strict life limits and continuous tracking of the fatigue life remaining for individual aircraft, using a quantifiable metric, FLE. Resource decisions are driven by the explicit structural-life limit and the FLE. Close working relationships and colocation facilitate communication.
- The Canadian Forces take a regulatory approach to structural-life management, having an independent regulatory authority that provides ASIP oversight and assesses compliance. The regulatory strategy allows ASIPs to be tailored for each aircraft type and allows case-by-case assessment of compliance. The regulations have a strong influence on communication processes and resource decisions.
- The U.S. Air Force has flexible, broad policies that allow ASIPs to be tailored to each MDS. Each SPO is responsible for implementing the ASIP for its MDS, but the lead command for that MDS is the primary authority for its ASIP because it holds the funding authority. There are minimal regulations and oversight. The U.S. Air Force lacks a standard measure that can be used to communicate the current fleet status effectively. The limited communications (partly because of geographic separa-

tion between the lead command and the SPDs and because of the lack of regulations) and the lead command's limited involvement in the ASIP process mean that the visibility of structural conditions may be inadequate for resource management.

Options for the Future

The Air Force has opportunities to enhance its ASIP by adopting and adapting some of the approaches the Canadian Forces and the U.S. Navy have taken. The following subsections suggest some options the Air Force may wish to consider.

Clear Policies

Clarifying ASIP policies and extending existing processes would enable independent assessment of ASIP compliance. Neither the AFPD nor the AFI currently mentions the ASIP standard, MIL-STD-1530B. Much as the Canadian Forces use an aircraft's basis of certification in determining its airworthiness, the Air Force could use compliance with MIL-STD-1530B as a measure of ASIP compliance and still allow certain degree of flexibility for tailoring. Periodic independent assessments of compliance, however, are necessary to ensure adequate levels of ASIP implementation.

The Air Force can build on existing processes to incorporate independent review functions to minimize added complexity. It can also help minimize the additional resources associated with institutionalizing a centralized regulatory body for ASIP. For instance, the U.S. Air Force's Fleet Viability Board (AFFVB) currently provides independent technical assessments of selected fleets.[1] The Air Force could expand the role of the AFFVB to include a review of each fleet's ASIP.

---

[1] The senior ASIP advisor (ASC/EN) is a member of the AFFVB.

Formal Processes and Independent Assessments

Formalizing key ASIP processes and assigning an independent assessment authority would ensure continued enforcement of ASIP and enhance communications. Formalizing a process involves a documentation, formal review, and approval process, either by an authority or by consensus of participants. Key ASIP processes, such as updating ASIP master plans and FSMPs, should be formalized to ensure that they are carried out adequately and meet the applicable ASIP standards. These processes should involve all affected organizations, including the lead command or its ASIP OPR. Involvement facilitates communication of the status of both ASIP and the fleets being managed, enhancing informed decisionmaking about ASIP funding. Such formal processes would strengthen communications between the lead command and the SPO.

To establish checks and balances for these processes, the Air Force could assign an independent assessment authority, similar to the organizational responsibility NAVAIR Air Vehicles has. The ALC/ENs at each center could provide independent technical assessments for the ASIPs housed within their ALCs. This would also promote ALC-wide communications among ASIP managers, providing visibility and cross-fertilization across fleets.

Good Communications and Working Relationships

Establishing close working relationships facilitates informal communication. To compensate for the geographic dispersion of its ASIP authorities, the Air Force could adopt the approach AMC and AFSOC have taken: assigning an engineer as OPR to act as a liaison with the lead command for technical issues and with the SPDs for operational issues. Such communication enables the lead commands to gain a holistic view of their fleet-management issues.

Additionally, the Air Force can take advantage of the colocation of the SPDs and ASIP managers at each ALC and implement periodic ALC-wide meetings to communicate ASIP-related issues and discuss potential solutions.

Standardization

Standardization would make commandwide visibility possible. Because the fleets within a command have commonalities in missions and aircraft use, some standardization of the ASIPs within a command may be achievable without much compromise in accounting for ASIP variability between the fleets. For example, standardizing data collection and data analysis within a command would help in developing a standard metric for structural condition that would help the lead command allocate resources. Centralizing these tasks and standardizing the processes might also cost the command less than would allocating funds for separate data collection and analysis tasks for each fleet using different methods.

As the Air Force develops its risk-assessment framework, which could provide a standard metric for lead commands to use in fleet management, the Air Force should clearly establish acceptable and unacceptable risk levels so that the decisionmakers clearly understand the implications.[2]

Dedicated Funding

ASIP tasks should have stable funding. Dedicating separate funding for tasks that are critical to ASIP viability would promote their stability. For instance, separate funding could be provided for regular FSMP updates to ensure that sufficient and accurate information about the fleet's structural condition is available. The funds would be dedicated to data collection (IAT, L/ESS, and maintenance records) and data analysis, as well as to evaluation and updates of fatigue analyses to ensure that the information in the FSMP is up to date and complete.

Before implementing any new processes, the Air Force must understand how its current policies and practices would complement or conflict with such changes. We recommend a deeper analysis of these suggested options to obtain a full understanding of their implications for the Air Force.

---

[2] MIL-STD-1530C in fact defines these risk levels.

History of the Air Force ASIP

The Air Force established its ASIP following a series of catastrophic structural failures of B-47 aircraft wings in 1958. These structural failures were caused by an unanticipated structural failure mode, fatigue. (Figure A.1 presents a timeline of dates affecting ASIP creation and evolution.) At the time of these accidents, the structural failure modes of fatigue were not well understood and had not been considered in structural design. Before these incidents, the Air Force had relied on a strength-based design concept, in which the aircraft structure was designed to prevent structural failure resulting from a single application of a load caused by an event (e.g., landing).

Figure A.1
Timeline of ASIP Evolution

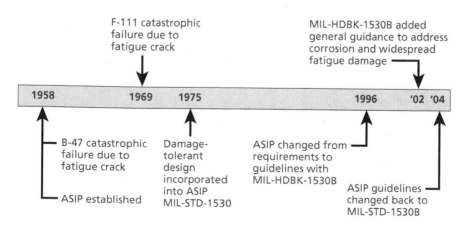

When the series of accidents demonstrated that the strength-based structural design philosophy was deficient for sustaining structural safety, the Air Force incorporated a fatigue-based design concept into its ASIP to prevent failures due to fatigue cracks during an aircraft's service life.

Then, in 1969, an F-111 crashed because of a fatigue crack after less than 100 hours of flight. The accident investigation revealed that an undetected manufacturing flaw in the wing pivot fitting had turned into a fatigue crack that grew rapidly, leading to catastrophic failure even though the crack was relatively small at that point (Tiffany et al., 1997). The Air Force was using the safe-life fatigue design approach at the time to determine structural design service lives. In the Air Force's safe-life approach, full-scale fatigue tests were conducted to determine the mean time to structural failure. The aircraft's service life was then determined by dividing this mean life by a life-reduction factor. The full-scale fatigue tests and a life-reduction factor of four that the Air Force was using at the time, however, did not prevent structural failure from such an anomaly.

The F-111 incident prompted the Air Force to examine whether ASIP sufficed to prevent catastrophic failures that were due to initial flaws in an element critical for flight safety. Starting in 1972, the U.S. Air Force therefore adapted its ASIP approach to incorporate damage-tolerance design concepts to protect against premature failures due to undetected defects that might have been introduced during manufacturing or that developed in service. In 1975, the damage-tolerance approach was incorporated into the ASIP standard, MIL-STD-1530.

ASIP continued to implement the damage-tolerance approach throughout the 1970s and the 1980s, augmenting it further with advanced analytical tools. In 1996, as part of the overall government effort to minimize acquisition regulations, the Air Force reduced the status of AFR 80-13, which governed ASIP, to that of an instruction and the status of MIL-STD-1530 to that of a military handbook (MIL-HDBK-1530B). The ASIP standard was thus no longer a set of requirements but guidelines.

In 2002, to address some of the issues for sustainment of aging airframes, the ASIP handbook was updated to include general guidance on other structural damage mechanisms (corrosion and widespread fatigue damage). Then, in February 2004, ASC/EN worked with SAF/AQ to convert MIL-HDBK-1530B back into a military standard that could once again be used as a requirement to reestablish some standardization of the ASIP.

Bibliography

Aging Aircraft Technologies Team (AATT), "USAF Aging Aircraft Structures and Mechanical Subsystems," ASC/AAA, June 2002. For Official Use Only.

Air Force Association, "USAF Almanac 2004," *Air Force Magazine*, Vol. 87, No. 5, May 2004. http://www.afa.org/magazine/May2004/default.asp (as of October 5, 2005).

Air Force Research Laboratory, Handbook for Damage Tolerance Design, Web site, 2002. Online at http://www.dtdesh.wpafb.af.mil/handbook.asp?url=contributors.htm (as of September 30, 2005).

Broek, David, *Elementary Engineering Fracture Mechanics*, Boston: Martinus Nijhof Publishers, 1986.

———, *The Practical Use of Fracture Mechanics*, Boston: Kluwer Academic Publishers, 1988.

Canada Department of National Defence, *Technical Airworthiness Manual*, October, 2003.

Dowling, Norman E., *Mechanical Behavior of Materials*, Englewood Cliffs, N.J.: Prentice-Hall, 1993.

Gallagher, J. P., F. J. Giessler, A. P. Berens, and R. M. Eagle, Jr., *USAF Damage Tolerant Design Handbook: Guidelines for the Analysis and Design of Damage Tolerant Aircraft Structures*, Wright-Patterson Air Force Base, Ohio: Air Force Wright Aeronautical Laboratories, AFWAL-TR-3073, 1984.

Grandt, Alten F., Jr., *Fundamentals of Structural Integrity: Damage Tolerant Design and Nondestructive Evaluation*, New York: John Wiley and Sons Inc., 2003.

Hebert, Adam J., "When Aircraft Get Old," *Air Force Magazine*, Vol. 86, No. 1, January 2003.

Jumper, John P., "The Future Air Force," remarks at the Air Force Association Air Warfare Symposium, Orlando, Fla., February 2003. Online at http://www.findarticles.com/p/articles/mi_m0PDU/is_2003_Feb_13/ai_107121397 (as of November 17, 2005).

Keating, Edward G., and Matthew Dixon, *Investigating Optimal Replacement of Aging Air Force Systems*, Santa Monica, Calif.: RAND Corporation, MR-1763-AF, 2003.

National Defence Canada, Canada's Air Force, Web Site, June 22, 2005. Online at http://www.airforce.forces.gc.ca/today5_e.asp (as of September 30, 2005).

NAVAIRINST—*See* Naval Air Systems Command.

Naval Air Systems Command, "Program/Project Risk Management," Instruction 5000.21.

———, "Fixed Wing Aircraft Structural Life Limits," Instruction 13120.1.

———, "Procedures for Submitting Flight Loads, Launch, and Landing Data for the Structural Appraisal of Fatigue Effects Program," Instruction 13920.1.

Office of the Chief of Naval Operations, "Policies and Peacetime Planning Factors Governing the Use of Naval Aircraft," Instruction 3110.11.

———, "Management of Naval Aircraft Inventory," Instruction 5442.8

OPNAVINST—*See* Office of the Chief of Naval Operations.

Pyles, Raymond A., *Aging Aircraft: USAF Workload and Material Consumption Life Cycle Patterns*, Santa Monica, Calif.: RAND Corporation, MR-1641-AF, 2003.

Rusk, David T., and Paul C. Hoffman, "Developments in Probability-Based Strain-Life Analysis," Aging Aircraft 2001: Conference Proceedings, Kissimmee, Fla., September 10–13, 2001. Online at http://www.jcaa.us/AA_Conference_2001/Papers/4B_2.pdf (as of October 5, 2005).

Secretary of the Air Force, "Report of the Secretary of the Air Force," in *Annual Report to the President and the Congress*, 2002. Online at http://www.defenselink.mil/execsec/adr2002/html_files/af_rpt.htm (as of October 5, 2005).

Shah, Bharat M., "P-3C Service Life Management," *Proceedings of 2004 USAF ASIP Conference*, Memphis, Tenn., November 30–December 2, 2004. Online at http://www.asipcon.com/media/proceedings/pdf%20 files/Tues_1000_Brussat.pdf (as of November 17, 2005).

Tiffany, Charles, et al., *Aging of U.S. Air Force Aircraft: Final Report, Committee on Aging of U.S. Air Force Aircraft*, National Materials Advisory Board, Commission on Engineering and Technical Systems, National Research Council, Washington, D.C.: National Academies Press, 1997.

U.S. Air Force, "Aircraft Structural Integrity Program," Air Force Regulation 80-13, October 1984.

———, "Aircraft Structural Integrity," Air Force Policy Directive 63-10, November 1, 1997.

———, "Lead Operating Command Weapon Systems Management," Air Force Policy Directive 10-9, June 13, 2000.

———, "Aircraft Structural Integrity Program," Air Force Instruction 63-1001, April 18, 2002.

———, "Aircraft Structural Integrity Program," Air Force Instruction 63-1001, April 7, 2003.

U.S. Department of Defense, "Standard Practice for System Safety," Military Standard 882D, January 19, 1993.

———, "Aircraft Structural Integrity Program," Military Handbook 1530B, July 3, 2002.

———, "Aircraft Structural Integrity Program," Military Standard 1530B, February 20, 2004.

———, "Aircraft Structural Integrity Program," Military Standard 1530C, November 2005.

U.S. Navy, Fact File, Web site, March 1, 2005. http://www.chinfo. navy.mil/navpalib/factfile/ffiletop.html (as of March 2005).